# Malabar Farm

# Malabar Farm

## Louis Bromfield, Friends of the Land, and the Rise of Sustainable Agriculture

Anneliese Abbott

The Kent State University Press KENT, OHIO

ISBN 978-1-60635-431-5
Manufactured in the United States of America



Cataloging information for this title is available at the Library of Congress.

25 24 23 22 21 5 4 3 2 1

In loving memory of Daddy

Steven Lee Abbott

December 30, 1960–January 6, 2020

# Contents

# Preface

I've been interested in the history of the modern American food system for a long time. Growing up in rural Michigan and helping out on my family's small farm gave me a deep appreciation for agriculture and farmers. I always knew what modern production agriculture looked like, and I knew how to milk goats and grow vegetables on a small scale, too. At the same time, I was intrigued by the past of American agriculture. I toured agricultural museums, read books like *Farmer Boy* by Laura Ingalls Wilder, and dreamed about how cool it would have been to live on a self-sufficient farm in the nineteenth century. But something didn't add up—if the agriculture they had back then was so great, why was it replaced with the modern system? How did American farming become so dependent on fossil fuels and chemicals? Why did organic and conventional farmers argue so much about what was sustainable?

When I started at the Ohio State University in 2013, I enrolled in the Sustainable Plant Systems major to learn how to grow food more sustainably. After discovering that OSU didn't offer a class on the history of agriculture, I asked the curator of the Rare Books and Manuscripts Library, Geoffrey Smith, if he had any special collections on agricultural history. "There's the Louis Bromfield Collection," he replied. I didn't know who Louis Bromfield was, but over the next year I checked out a couple of his books from the library and was intrigued. Bromfield, I discovered, had established Malabar Farm in 1939, practicing something called *soil conservation* to restore his worn-out land.

Two things about Malabar Farm were especially eye-opening for me. First, Bromfield was critical of some of the pioneer agricultural methods that I had romanticized; he said that things like the moldboard plow, even if drawn by horses, were inherently unsustainable. Second, most of the soil conservation practices Bromfield talked about sounded a lot like sustainable agriculture. He talked about many things that I had thought were new: conservation tillage, the dangers of pesticides, grass farming, the organic versus chemical fertilizer controversy, and even how to feed the world. I was learning about those same issues in my classes, and I was amazed by how little the discussions about them had changed in seventy years. Bromfield also helped me understand the importance of soil. I had always thought plants were the most important part of the food system, since animals and humans are dependent on them, but I hadn't really thought about how important soil was for growing healthy plants. I decided to minor in soil science to learn more about this fascinating topic.

As an honors student, I had to complete a research project in order to graduate. A plant science student would typically have worked in one of the labs on campus, but Jim Metzger, my advisor and professor and chair of the Department of Horticulture and Crop Science, gave me permission to do a library research project instead and get into the Louis Bromfield Collection. Geoff Smith, Rebecca Jewett, and the friendly student employees in the Rare Books reading room helped me get started on the archival research, and I spent much of the spring 2015 semester going through some really fascinating archival documents. I visited Malabar Farm for the first time in the summer of 2015, where park manager Korre Boyer and unofficial Malabar historian Tom Bachelder spent an entire day giving me a behind-the-scenes tour. I finished my research and wrote an honors thesis in time to graduate in spring 2016.

About a year later, I contacted Will Underwood at Kent State University Press about possibly publishing my thesis. An anonymous reviewer for the Press read my proposal and recommended I do a lot more background research to put Malabar Farm in its historical context. Over the next two years, I read hundreds of primary and secondary sources, both books and articles. I revisited the archives at OSU and the Ohio History Connection, where the friendly staff very generously helped me find all the materials I needed. When I returned to Malabar in 2018, Tom Bachelder helped me look at archival documents in the Big House and took me on a hike to see all of Malabar's caves.

Kim Craven and Brenda Brownlee, librarians at the Noble Research Institute in Ardmore, Oklahoma, scanned hundreds of pages of archival documents about Malabar for me. Former park managers Jim Berry, Brent Charette, and Louis Andres and current park manager Jennifer Roar shared their experiences at Malabar with me. Scott Williams, Mick Luber, and Jodine Grundy helped fill in some of the gaps about the Ohio Ecological Food and Farm Association's partnership with Malabar and the birth of sustainable agriculture in the 1980s.

Fortunately, I had already accessed all the archival and print resources I needed before all the libraries shut down due to the COVID-19 pandemic in March 2020. By April, I was ready to begin writing, and I had many wonderful people review the manuscript for me: Joseph Heckman, professor of soil science at Rutgers University; Peg McMahon, Associate Professor Emeritus in the Department of Horticulture and Crop Science at OSU; Susan Kline, associate professor in the School of Communication at OSU; Thomas Bachelder, Jim Berry, Brent Charette, and Louis Andres. David Barker, professor and associate chair of the Department of Horticulture and Crop Science at OSU, reviewed the chapter on grass farming. Nan Enstad, professor in the Department of Community and Environmental Sociology at the University of Wisconsin–Madison, gave me some helpful criticism on the historical background in the first chapter.

I started graduate school in the Nelson Institute for Environmental Studies at the University of Wisconsin–Madison in fall 2020 and was privileged to be in Bill Cronon's last class on American environmental history. My final revision was influenced by ideas presented in this class, and the extensive bibliographies Bill provided for each lecture helped guide me to the relevant literature on several topics. Susan Wadsworth-Booth from Kent State University Press, two anonymous reviewers, and copyeditor Erin Holman also gave valuable feedback on the manuscript. The final decision about how to tell the story of Malabar and the viewpoints expressed in this book are my own. I have tried to verify all facts to the best of my ability and apologize for any errors that remain.

Throughout the long writing process, my parents, Steve and Caroline Abbott, and my siblings, Elizabeth, Laura, and Tim, were very supportive. Mom always let me talk to her about the things I was learning in my research and gave me lots of valuable feedback, and the critical thinking skills she taught me when I was a child have served me well throughout the research and writing process. Daddy was proud of my research work and used his frequent traveler points to get me hotel rooms while I was traveling in 2018, just before he was diagnosed with acute lymphoblastic leukemia that September. Even though at times it was hard, I am grateful that I was able to spend the next year and a half with him, before he passed away on January 6, 2020. I am very thankful to have been blessed with such a wonderful family.

# Introduction

Malabar Farm is a unique part of the Ohio state park system. There isn't anything in the United States exactly like this thousand-acre park of rolling green fields and pastures, forested hills, and historic buildings, about ten miles southeast of Mansfield, Ohio. Like most other state parks, Malabar Farm has lots of recreational opportunities, including three nature trails that lead through shady forests, beautiful wetlands, and even a rock crevice "cave" in the Blackhand sandstone. The park has a working farm, too, featuring beef cattle that spend at least part of the year on pasture and live the rest of the time in a historic barn on Bromfield Road. Beef from the Malabar herd is sold in the gift shop and served along with other local produce at the Malabar Restaurant, located in a beautifully restored 1820 brick building, the oldest structure in the park.

Malabar Farm is located in north-central Ohio, near Mansfield. (Map by author)

What really makes Malabar Farm special is the park's history. In 1939, the world-famous novelist Louis Bromfield moved back to his home state of Ohio and constructed the one-of-a-kind thirty-two room Big House, which has been carefully preserved as a time capsule of life in the 1940s. The Bromfield family's historic furniture—including some seventeenth-century French pieces—is still arranged as it was seventy years ago. Twin red-carpeted staircases curve up to the second floor in the front entryway; the parlor features a mirrored wall decorated with a gold eagle and stars; and the walls are still

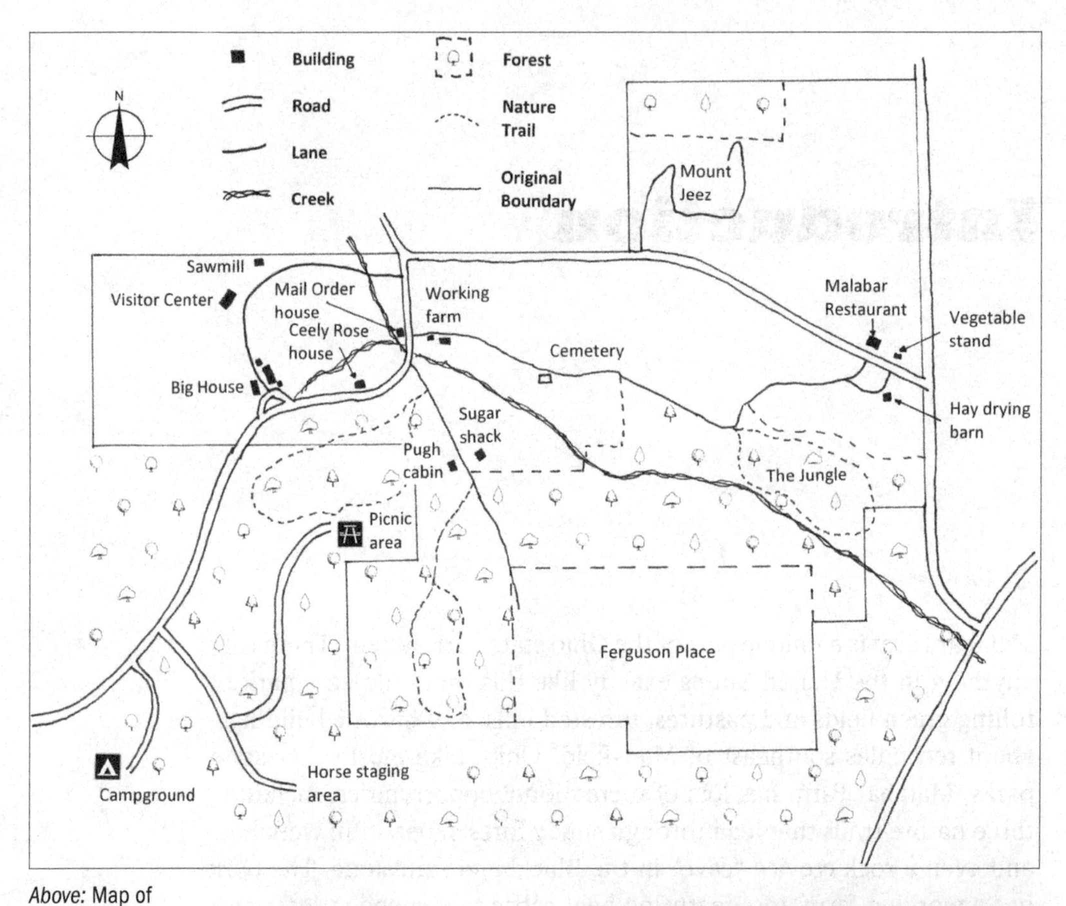


*Above:* Map of Malabar Farm, showing major locations referred to in this book. (Map by author)

*Right:* The Big House. (Photo by author)

The "haunted" Ceely Rose house. (Photo by author)

covered with the remarkably preserved original wallpaper, featuring custom designs such as the raspberry and clamshell pattern in the sitting room. Visitors can tour most of the house, including Louis Bromfield's combination study and bedroom; a wood-paneled room on the second floor where his secretary, George Hawkins, lived; and the guest bedroom where movie stars Humphrey Bogart and Lauren Bacall spent their wedding night after being married at Malabar in 1945.

Throughout the year, Malabar Farm State Park hosts activities for the whole family, including the Maple Syrup Festival in March and Ohio Heritage Days, the state's largest outdoor history festival, in September. Visitors can drive or walk to the top of Mount Jeez, the highest hill in the park, to see a panoramic view of the farm's hills, fields, and buildings. Kids can see and pet goats and other farm animals in the barn across from the Big House, and the park offers tractor-drawn wagon tours. Highlights on the farm tours include an aluminum-sided barn where Bromfield once tried to dry hay, just across the street from the Malabar Restaurant; a sandstone vegetable stand, where he marketed fresh vegetables cooled in spring water; a small cemetery where Louis Bromfield's ashes are buried; and a nondescript little white house on the banks of Switzer's Creek that some people claim is haunted by the ghost of Ceely Rose, a mentally ill girl who murdered her entire family there long before Bromfield bought the farm.

Many visitors to Malabar Farm may not realize, however, that Malabar's history is about far more than a famous novelist, Hollywood weddings, and ghost stories. Here, fifty years before the phrase *sustainable agriculture* entered the American vernacular, Louis Bromfield was promoting a system that looked a lot like it. The 1930s, when he purchased the property, were a time of crisis in American agriculture. Farmers struggled financially, poor farming practices eroded soil and depressed yields, and many believed that the future not only of American agriculture but of the nation itself was in jeopardy. During the New Deal era, out of this turmoil rose the largest movement for soil conservation in American history. Ecologists and soil scientists worked together to develop farming systems that stopped soil erosion and increased yields and profitability at the same time. Louis Bromfield established Malabar Farm at the height of this movement, and soon he was at the cutting edge, experimenting with and promoting novel techniques like conservation tillage, grass farming, and direct marketing pesticide-free vegetables. Each year, thousands of visitors came from all over the world to see and learn from what was happening at Malabar, which soon became known as "the most famous farm in the world."[1]

One reason Malabar Farm's agricultural legacy is not well-known today is that the soil conservation movement that made the farm famous was almost completely left out of agricultural history books published during the Cold War. Instead, agricultural historians focused on the economic, political, and technological factors that shaped the modern food system. The story they told was one of positive, inevitable evolution from rudimentary pioneer methods to the sophisticated technology that allows less than 1 percent of the American population to grow food for everyone else today. In this narrative, environmental factors in general were almost completely neglected, and the soil conservation movement was usually glossed over as if it had never happened. The notion of working within natural laws to develop a permanently sustainable agricultural system didn't fit well with the popular 1950s belief that humans could control nature, and so it was easiest to ignore soil conservation altogether.[2]

The environmental movement, which exploded in the 1970s, questioned whether the march of technology was inevitable or even positive. Environmentalists wrote their own versions of American history, emphasizing the exploitation and destruction of natural resources, especially forests and wildlife. In this narrative, technology and industrialized agriculture were portrayed as negative, increasing in destructive power as they increased in technological sophistication. Central to the classic environmental narrative was the concept of wilderness—untouched, primeval nature that would be destroyed by any human intervention. Despite its emphasis on ecology, soil conservation didn't fit well into the 1970s environmental narrative, mostly because environmentalists focused almost exclusively on preserving wilderness areas and fighting pollution in cities. Agricultural areas like Malabar Farm, which were where most of the soil conservation movement was focused, had no place in this urban/wilderness dualistic view of nature. As the pioneering environmental historian Roderick Nash put it, "wilderness is just as dead in the garden as it is in the concrete wasteland."[3]

"The wilderness dualism tends to cast away any use as *ab*-use, and thereby denies us a middle ground in which responsible use and non-use might attain some kind of balanced, sustainable relationship," environmental historian William Cronon warned in 1995. In the 1990s, environmental historians began calling for a broader version of environmental history with more focus on human-modified ecosystems, including agricultural land and food systems. Since then, the "second generation" of environmental historians has begun to rediscover the New Deal conservation movement and its emphasis on sustainable agriculture and soil conservation, mostly as part of an increasing body of work on southern environmental history. In the past thirty years, there has been an increasing awareness that agricultural systems can vary widely in their impact on the environment. Some are exploitive and rapidly degrade the soil; others have the potential of being sustainable for hundreds or even thousands of years. In light of this renewed interest in the connection between agriculture and the environment, it is time to put Malabar Farm back into its rightful place in the history of American agriculture.[4]

Malabar Farm is in that often overlooked but critically important middle ground between the primeval wilderness and the concrete city—the rural landscapes where most of the food, shelter, and clothing for the rest of the country are produced. About 50 percent of the total land in the United States is used as cropland, pasture, or rangeland, and another 20 percent or so is used for forestry—far more than all the cities and wilderness areas combined. Only 16 percent of all Americans live in these rural areas, and less than 1 percent of the population of the United States is directly engaged in farming, forestry, and other uses of the renewable natural resources on this rural land. While wilderness areas are valuable for recreation and preservation of biodiversity, and cities are important centers of culture and socialization, neither could be enjoyed by the general population without a steady supply of food and other natural resources from the agricultural rural areas. Yet the majority of Americans know or care little about how these vital rural lands are managed.[5]

"There will still be a farmer—after there is no city—because when the soil is so depleted that it produces only for one, it will be the farmer who lives and the city man who starves," the soil conservation organization Friends of the Land wrote in 1956. "Conservation permits the farmer to produce both his own needs—and then there are farm products to sell to those who live off the land in cities. Therefore, in the period of survival, conservation is of greater importance to those of us who live in cities than it is to those who live on the farms." Today, both urban and rural Americans are still completely dependent on farmers to grow their food. It may be processed beyond recognition and trucked thousands of miles across the country, but except for a small amount of hydroponic specialty crops and seafood, almost every bite we eat still comes from the soil. Everyone, whether they realize it or not, still has just as much of an interest in making sure that soil is managed sustainably as they did in Louis Bromfield's day.[6]

As agronomists, agricultural economists, and environmentalists continue to debate over how to make American agriculture more sustainable, perhaps the story of Malabar Farm can help fill a major gap in the current dialogue. As one of the few surviving landmarks of the soil conservation movement, Malabar Farm provides an intriguing case study of how that movement began, how it was marginalized during the 1950s, and how it influenced the ideas that collectively became known as "sustainable agriculture" in the late 1980s. The story of Malabar begins at the height of the New Deal soil conservation movement, when visionary pioneers dreamed of a permanent agriculture that worked with nature instead of against it. It traces what happened to that dream—embodied in Malabar Farm—through years of unprecedented changes, paradigm shifts, and conflicts, and how it managed to survive despite what at times seemed like overwhelming odds. It provides a much-needed historical background to the current discourse on sustainability, a deeper perspective on the factors that have influenced the rise and fall of sustainable farming systems in the United States over the years.

It's time to rediscover why Malabar Farm was called the most famous farm in the world.

*Chapter 1*

# Louis Bromfield Comes Home

In the fading light of a clear winter's day in 1938, the world-famous author Louis Bromfield knocked on a farmhouse door in northcentral Ohio. Accompanied by his wife, Mary, and his secretary, George Hawkins, Bromfield had been driving along narrow, winding country roads for most of the afternoon. "It was already twilight and the lower Valley was the ice-blue color of a shadowed winter landscape at dusk and the black, bare trees on the ridge tops were tinted with the last pink light of the winter sunset," Bromfield reminisced several years later. He had spent his childhood in this valley, and the sight of each hill and stream stirred up memories from the happy days of youth. But this was no mere nostalgic trip; Bromfield intended to settle his family here, far from the war that would soon engulf Europe. He finally stopped in front of an especially attractive farm, with the farmhouse perched halfway up a picturesque hill. "Like Brigham Young on the sight of the vast valley of Great Salt Lake, I thought, 'This is the place,'" Bromfield said. Soon the door of the farmhouse opened, and out stepped the farmer's wife to greet Bromfield. Behind her, a delicious fragrance wafted out from the farmhouse kitchen—a combination of wood smoke, apple butter, sausage, and pancakes. The smell brought back memories of his childhood, sharply and vividly—and Louis Bromfield knew that he had come home.[1]

Forty years earlier, on December 27, 1896, Louis Bromfield had been born in the city of Mansfield, Ohio. His parents, Charles and Annette Brumfield, regarded his birth as nothing short of a miracle. After Annette had nearly died giving birth to her first son, the doctor told her that she would be unable to bear more children. When her second son was born healthy, Annette christened him "Lewis Brucker Brumfield"; he would later change the spelling of his name to "Louis Bromfield." According to

The farmhouse Louis Bromfield saw that winter evening. (Malabar Farm archives, courtesy of Ohio Department of Natural Resources)

Louis, before he was even born his strong-willed mother decided that her son should be a writer. She purchased an entire classical library and encouraged him to read and write as much as possible. Whether due to his mother's "pre-natal influence," Louis Bromfield indeed developed a love of writing and literature that stayed with him his entire life.[2]

Bromfield's love for literature was matched only by his love for the land. His own family lived in town, but every Sunday they would harness up the horses and head out to his grandparents' farm on the outskirts of Mansfield. This land had been in his mother's family since the late eighteenth century, when the first pioneers settled Ohio. Young Louis loved everything about this farm, especially the food—fresh milk and cream from the cows, chicken and sausage, home-grown fruits and vegetables, pies, and maple syrup. He enrolled at Cornell Agricultural College in the fall of 1914 to study agriculture, intending to return to the family farm in Ohio after graduation. But he had completed only one semester at Cornell when he received news that his grandfather had broken his hip and could no longer run the farm. Louis returned to Mansfield and managed the farm, "not too badly," for a year. He quickly realized that he was too restless to be a farmer; he could not endure the thought of spending his whole life tied down to one piece of land. When his grandfather died and the farm began to struggle financially, Bromfield decided that writing was a wiser career choice. In 1916, the ancestral farm was sold, and Louis Bromfield left for the Columbia University School of Journalism.[3]

Ever restless, Bromfield never graduated from college. Before the end of his first year at Columbia, he left for France to serve as an ambulance driver in World War I. He stayed in France for a while after the war ended and then returned to the United States, where he began his writing career as a reporter in New York City. In New York, Bromfield fell in love with Mary Appleton Wood, a quiet woman from an upper-class New England family, and they were married in 1921. After working several other writing jobs, Bromfield published his first novel, *The Green Bay Tree,* in 1924. Set in a midwestern industrial town that was a fictionalized version of Mansfield, *The Green Bay Tree* was favorably received by readers and reviewers alike and launched Bromfield's career as a bestselling author. Louis Bromfield wrote more than thirty books in his lifetime, turning out nearly a novel a year during the 1920s, and quickly became one of the wealthiest and most famous American authors of the time. In 1927, he received the Pulitzer Prize for his 1926 novel *Early Autumn,* set in New England. One of his most famous books was *The Rains Came* (1937), about contemporary India, but many critics regarded his 1933 fictionalized autobiography, *The Farm,* as Bromfield's best novel ever.[4]

Louis and Mary Bromfield had three daughters: Anne, Hope, and Ellen. For over a decade, the Bromfield family made Senlis, France, their home base, though Louis Bromfield often traveled to England, New York, Hollywood, Switzerland, and India, sometimes taking Mary or his secretary, George Hawkins. Everyone considered Hawkins, who edited Bromfield's most successful novels, a member of the family. Jean White, the girls' nanny when they were young, was also an integral part of the Bromfield household. In Senlis, the Bromfields lived in the Presbytére de St. Etienne, an old house with a rich history that they leased from three elderly ladies. Louis Bromfield was happy in Senlis and might have stayed there for the rest of his life, if not for the impending threat of World War II. Louis sent Mary and the children back to America but was reluctant to leave France himself. Finally his good friend Louis Gillet urged him to return home. "There is nothing you can do here," Gillet urged. "Go home and tell your people. You can help France most by doing just that."[5]

So, after wandering the world and gaining fame and wealth, Louis Bromfield returned home, to the land he had known as a child, to Pleasant Valley. In 1939, he purchased three adjoining farms, totaling 355 acres—the Herring farm, where he had knocked on the door on that snowy evening; the adjacent Beck farm; and the high and lonely Ferguson farm. He added a marshy seventy-acre triangle of land called "The Jungle" in 1940 and a fourth farm, purchased from Laura Niman, in 1942, bringing the total to six hundred acres. Bromfield christened the combined property Malabar Farm, after the beautiful southwest coast

Bromfield family on a ship, 1932. *From left:* Louis, Anne, Hope, Mary, George Hawkins, Jean White. (Malabar Farm archives, courtesy of Ohio Department of Natural Resources)

of India where he had spent several happy months doing research for his popular novel *The Rains Came.*[6]

Using the old Herring farmhouse as a nucleus, Bromfield commissioned a Mansfield architect named Louis Lamoreux to construct a one-of-a-kind dwelling called the "Big House," which took a year and a half to build. Bromfield wanted the house to look like it had been added onto over many years, so it contained elements of architectural styles from many different periods of Ohio history. He found local craftsmen who were skilled woodworkers and could make every detail look authentic. The rooms were large, brightly colored, and designed specifically for each member of the family. Bromfield spared no expense in building the house, but it was meant to be lived in. As Lamoreux put it, "With the aid of his dogs, many guests, children and the continual flow of activity, the place soon looked as if it had always been there."[7]

Part of the reason Bromfield put so much time and care into the Big House was that he intended it to be his fortress. When he returned to Ohio, he was in a sense running away from the war and turmoil that was shaking the entire world. He wanted Malabar to be an old-fashioned "general" type of farm where everything that was served at the table had been grown in the fields outside. With the threat of war on the horizon, he foresaw rationing and possibly even food shortages. Bromfield wanted Malabar to produce practically everything: chicken, eggs, milk, butter, beef, pork, fruits, vegetables, assorted poultry, fish, maple syrup, honey, and more. He believed that the farm could support four or five families comfortably and lavishly, under a management system that he called "the Plan." Bromfield's Plan was modeled after the Russian collective farm, with Bromfield serving the role of the state. He would provide housing, electricity, heat, produce from the farm, and a salary to the workers and their families, and they would farm the land. Bromfield envisioned Malabar as a self-sufficient, secure, peaceful "island of security" that could "withstand a siege and where, if necessary, one could get out the rifle and shotgun for defense."[8]

Bromfield's utopian vision of a productive, self-sufficient farm received a cruel blow when spring came. "When the snow was gone, I discovered that the valley of my childhood was no longer there. Something had happened to it," he wrote later. Formerly fertile fields were barren and produced only scraggly, acid-loving vegetation. While some of the lower fields were still fairly productive, the steeper slopes suffered from severe soil erosion. Jagged gullies cut into the sides of the hill behind his new Big House—one was so large that Bromfield's newly hired farm manager, Max Drake, joked, "You could lose a horse in it, bury it and nobody would ever know it was there." Almost all of the original topsoil on the tallest hill in the region, which locals called "Poverty Nob," had washed down onto the lower fields, creating a "sickly, unnatural marsh." Later, Bromfield would find drainage tiles (usually installed just a couple feet deep) fifteen feet below the soil's surface. He even dug up an old bridge that had been buried under six feet of displaced topsoil. And the hills on neighboring farms looked just as bad or even worse. In 1937, a reconnaissance survey of soil erosion in Ohio discovered that three-fourths of Ohio's farmland was affected by soil erosion, 12 million acres had lost more than a quarter of their topsoil, and 1.6 million acres had "been rendered useless for farming." In Richland County, where Malabar was located, most of the land had lost 25 to 75 percent of its original topsoil, with a gully every hundred feet or so on hilly land.[9]

The tragic story of what had happened to the topsoil started with the first arrival of European settlers in the New World. Early explorers brought back reports of a seemingly endless forest that covered half the North American continent, filled with an "inexhaustible" abundance of wildlife and other natural resources. To European observers, the Native Americans seemed to be ignoring or even wasting the land's potential wealth. The reality was much more complicated; humans had modified the American landscape more than the colonists acknowledged. Many indigenous nations in the eastern US had extensive villages and settlements, including agricultural fields that were sometimes hundreds of acres in extent. Because they relied on hunting for meat instead of raising livestock, native peoples often managed portions of the forest using controlled burns and other methods to provide forage and habitat for game animals. But diseases accidentally introduced by European traders and explorers had drastically reduced indigenous populations, and most settlers did not recognize the role native management practices played in maintaining the abundance of wildlife and other natural resources they saw on the North American continent.[10]

Colonists considered it their first order of business to "civilize" this wild land by trying to replicate the agricultural landscapes of Europe. Thus began "a transformation unlike any in the history of western civilization." The forests fell to axe and saw, and the soils beneath them were torn by the plow and seeded with various crops—either European staples, like wheat, or crops adopted from the Native Americans, like corn and tobacco. The pioneers were extremely wasteful as they converted the forest into farmland. An estimated 90 percent of the continent's original virgin stands of hardwood timber were clear-cut to make room for agriculture, and since there was no way to transport the wood to market, any of the enormous trees not used to build cabins or fences were rolled into huge piles and burned, "a vast initial sacrifice to progress."[11]

With a few exceptions—most notably the Dutch and German settlers in New York and Pennsylvania—pioneers practiced an inherently exploitative form of agriculture. While some lauded the pioneers as industrious, self-sufficient, and ingenious, others noted that the individuals who migrated to the frontier were often the outcasts of society and knew little about proper soil management for long-term, sustainable farming. Usually the newly cleared land produced phenomenal harvests for several years, but then yields started to decline. As early as 1747, some formerly productive land in the eastern United States could only grow turnips the size of buttons. At least one European visitor called the colonists "the worst slovens in Christendom" because they were such poor stewards of their land. But these criticisms did little to change American agriculture. Instead, the federal government continued to dispossess the Native Americans, opening up more "virgin" land farther west for the pioneers to settle. When yields started to decline, rather than try to rebuild their soil's fertility, many settlers just packed up and moved west.[12]

Many of the Founders, including George Washington, Thomas Jefferson, Benjamin Franklin, James Madison, and John Adams, were concerned with the slovenly state of American agriculture. On their own land, they practiced good agricultural practices such as crop rotation, manuring, contour plowing, and terracing hilly fields. During the 1840s and 1850s, many farmers in the Northeast and Southeast practiced *agricultural improvement,* which emphasized on-farm nutrient cycling in imitation of natural systems. Unfortunately, soil health was often neglected when the focus of the agricultural improvement movement shifted toward yields and profits in the second half of the nineteenth century. In the aftermath of the Civil War, many farmers, especially in the South, found that growing monocultures of commodity crops like cotton or tobacco was the only way to market enough produce to earn a living. Recommended "best" practices to grow these crops were sometimes detrimental to soil health—for example, when farmers began to rely on newly available commercial fertilizers instead of manure and crop rotations to maintain soil fertility, the resultant depletion of soil organic matter made the soil more susceptible to erosion.[13]

Meanwhile, many Americans began to worry about the destruction of other natural resources, such as forests, wildlife, and waterways. One of the most influential critics of the wasteful way Americans were managing their once-rich inheritance was George Perkins Marsh, who published a seminal book entitled *Man and Nature* in 1864. "Man has too long forgotten that the earth was given to him for usufruct alone, not for consumption, still less for profligate waste," Marsh warned. If current trends continued, the earth would soon deteriorate "to such a condition of impoverished productiveness, of shattered surface, of climatic excess, as to threaten the deprivation, barbarianism, and perhaps even decline of the species." The only way to prevent such a decline, Marsh argued, was to preserve and wisely manage the remaining natural resources.[14]

Concern about the destruction of natural resources grew during the last few decades of the nineteenth century, with various groups campaigning to save forests, wildlife, and scenic natural areas like Yellowstone from exploitation. During the first two decades of the early twentieth century, these concerns coalesced into a new movement for the conservation of natural resources. *Conservation,* as Gifford Pinchot defined it, was "the wise use of the earth and its resources for the lasting good of men" and "the foresighted utilization, preservation, and/or renewal of forests, waters, lands, and minerals, for the greatest good of the greatest number for the longest time." Conservation soon became an integral part of the Progressive political movement, which believed that government and science, by working together, could solve any problem besetting the human race and create a permanent, prosperous civilization. Until the 1970s, *conservation* was the word used to refer to any type of environmental protection or management; it had a broad meaning that encompassed much of what modern terms like *environmentalism* and *sustainability* seek to describe.[15]

Two leading figures in the Progressive conservation movement were President Theodore Roosevelt and Gifford Pinchot (1865–1946), who became the head of the USDA's Forest Service in 1898 and worked with Roosevelt to form the National Forest system that still exists today. Pinchot's two passions were forestry and politics, and his combination of the two made him a controversial figure. Educated at the French Forestry School in Nancy, L'Ecole Nationale Forestiére, Pinchot was impressed by the scientific forestry practices that he saw in Europe. He returned home in 1890 with the vision to see America's forests managed responsibly, not under the cut-and-get-out mentality that was rapidly demolishing millions of acres of virgin timber. Pinchot believed that one reason sustainable forestry was not being practiced in the United States was that there were only two, opposing, viewpoints about the forest—that it should be logged as fast as possible and that it should be preserved forever and never logged. Under proper management, Pinchot argued, forests were a renewable resource that could produce a *sustained yield* of lumber. Though Pinchot believed that in most cases forests should be used, not left as untouched wilderness, he understood that there were a few special cases—such as the giant sequoias in California—where trees should be left to grow.[16]

Along with their emphasis on conserving natural resources like water, forests, wildlife, and minerals, Progressive conservationists and some agricultural scientists also stressed the importance of conserving the nation's soil resources. In Illinois, Cyril Hopkins (1866–1919) promoted a system called "permanent agriculture," which used lime, rock phosphate, green manure, and crop residues to restore worn-out soil. Franklin Hiram King (1848–1911), a Wisconsin soil scientist, traveled to China, Japan, and Korea and wrote a book called *Farmers of Forty Centuries* about how Asian cultures had maintained their soils at a high level of fertility for thousands of years by composting all organic wastes and returning them to the soil. At Tuskegee Institute in Alabama, George Washington Carver (1864?–1943) taught African American farmers to collect and compost leaves, manure, swamp muck, and other organic wastes to increase soil fertility and organic matter. First in Michigan and later in New York, horticulturist and country life advocate Liberty Hyde Bailey (1858–1954) emphasized that "the greatest of all resources that man can make or mar is the soil." Even urban businessmen, worried that unchecked soil erosion might eventually threaten their food supply, formed an organization called the National Soil Fertility League.[17]

Despite these efforts, soil conservation did not have a major impact on American farming practices during the Progressive Era. The USDA's official viewpoint, formulated by Bureau of Soils chief Milton Whitney (1860–1927), was that soil fertility was inexhaustible. Early in his soil science career, Whitney and his colleagues ran some chemical tests on soils and discovered that most soils had a similar mineral composition, containing nutrients far in excess of plant requirements. This led him to boldly state in 1909, "The soil is the one indestructible, immutable asset that the Nation possesses. It is the one resource that cannot be exhausted; that cannot be used up." Cyril Hopkins and other permanent agriculture advocates disagreed, fearing Whitney's ideas would cause "enormous and irreparable damage" by lulling people into a false sense of security. Their warnings went mostly unheeded, partly because Whitney censored all official USDA soil bulletins and partly because American farmers were in the midst of what would later be considered a "golden age" of prosperity. For one of the only times in American history, farmers had "parity" or equal purchasing power with urban citizens. Things were going great; why listen to prophets of doom?[18]

The problem was that all was not well with American soils, and no one knew it more than one of Whitney's soil surveyors, Hugh Hammond Bennett (1881–1960). Bennett grew up on a mostly self-sufficient farm in North Carolina, studied geology and chemistry at the University of North Carolina, and got a job as a soil surveyor in the Bureau of Soils in 1903. As Bennett traveled across the country studying soils, he noted that many cultivated fields were exhausted and gullied, while the soil under adjacent woodlands was still soft and rich. Bennett was one of the first to attribute this to *sheet erosion,* the process by which rainwater washed topsoil particles downhill. He called this "the least conspicuous and the most insidious type of erosion," because most farmers did not notice that they were losing their topsoil until yields started to drop or the color of the soil seemed to change when the lighter subsoil was exposed. By then, of course, the damage was done. Most farmers didn't notice that their fields were losing soil until erosion had progressed to the stage of forming *gullies,* huge runoff channels that became deeper and wider with each successive rainstorm. Left unchecked, these gullies soon became monstrous ravines that swallowed up tens of thousands of acres of fertile land.[19]

One of the most famous gullies in the United States was Providence Canyon in Stewart County, Georgia. Legend had it that the gully started "from the roof drip of a barn it since has swallowed" and spread to at least three thousand acres of farmland. Journalist Stuart Chase described it as "awful and beautiful" and "a sickening void." Looking down into the canyon, he observed that "the earth strata changed from red to yellow to brown, mauve, lavender, jade, ocher, orange and chalk white. Pinnacles rose from the gully floor, sometimes with a solitary pine tree on their top at the level of the old land. . . . Stewart County has gone. No work by man or nature can bring it back within a calculable future." Providence Canyon was cited as a worst-case example of extreme soil erosion, although the farmers in Stewart County were not necessarily more destructive than anyone else. The problem was that the clay subsoil in the area rested atop a hundred feet of "unconsolidated sand" that was "largely structureless." Once erosion channels breached the shallow clay layer and reached this sand formation, catastrophic soil erosion was the result. Even though they knew the underlying geology contributed to the problem, advocates of soil conservation couldn't resist using Providence Canyon as their posterchild for soil erosion because it was so shocking.[20]

While perhaps the most extreme, Providence Canyon was certainly not the only example of catastrophic soil erosion in the United States. In Silver City, New Mexico, late-nineteenth-century city planners made the mistake of platting out the town's Main Street along a runoff channel because it was the flattest ground around. Beginning in 1895, a series of massive floods swept down the street, widening and deepening the channel until it became known simply as "The Big Ditch." In 1902, raging floodwaters even swept a grand piano from the second floor of a brick building and carried it seven miles along the wash. Soil erosion in northcentral Ohio, though not quite as dramatic, was still causing serious damage to farmland. "A study of Ohio crop statistics for the past 50 years shows that in spite of the increased use of fertilizers and lime, better methods of tillage, improved varieties of crops, and the like, yields have been maintained at about the same level, but have not been increased," a group of researchers from the Ohio Agricultural Experiment Station wrote in 1937. "The only logical conclusion is that the soils are deteriorating, and that this is resulting in a decreasing productive capacity."[21]

Severe gully erosion, 1930s. Note horse and rider at edge of gully for scale. (Natural Resources Conservation Service)

For twenty years, Hugh Bennett quietly collected data to bolster his case for soil conservation, awaiting his chance to bring his findings to the public. That chance came in 1927, the year Milton Whitney retired, when the USDA appropriated $160,000 to establish ten erosion research stations across the country. In 1928, Bennett published the preliminary results of his erosion investigations in a disturbing and groundbreaking USDA bulletin, titled *Soil Erosion a National Menace*. Bennett listed case after case of severe water erosion impoverishing formerly fertile fields across the nation. He estimated that the United States was losing 1.5 billion tons of soil annually; later he would set the figure nearly twice as high. In addition to this official publication, Bennett wrote numerous articles for popular and scientific magazines to raise public awareness of the erosion problem. But it took one of the worst ecological disasters in American history—the Dust Bowl—to finally shake the nation out of its complacency and get the government firmly behind soil conservation.[22]

"I don't care who describes it to you, nobody can tell it any worse than what it was. And no one exaggerates; there is no way for it to be exaggerated. It was that bad," a Dust Bowl survivor named Don Wells later told historians for a documentary project. The stage for the Dust Bowl had been set in the late nineteenth century, when the Great Plains, the semiarid grassy region that stretched from Texas to the Dakotas, was opened for settlement. Most of the pioneers who settled in the Great Plains tried to farm with the same methods they had used in the well-watered eastern states, but they often met with failure because the average rainfall on the Plains even in a good year was twenty inches or less. One crop that did quite well in wetter years was wheat, and when wheat prices soared in the late 1920s, farmers began to plow up the prairie grasses to plant wheat on a large scale. By 1931, over 30 million acres of shortgrass prairie in the southern Great Plains had been converted to wheat production. The bumper crop of 1931 flooded the markets with an immense surplus of wheat, and prices plunged to a mere twenty-five cents per bushel, only half of the farmer's production cost. With payments overdue on expensive farm machinery, farmers believed their only chance of breaking even was to plow up even more land in 1932.[23]

But then a series of dry years, with too little rainfall to support wheat production, struck the Great Plains. Farmers plowed their land and planted their wheat, but there was not enough moisture in the soil for crop plants to grow. Unable to pay their bills, farmers lost their savings, machinery, and sometimes even their homes. But that was not the worst. With no water and no vegetation, those 30 million acres of bare soil soon began to blow in the wind, causing massive dust storms that buried gardens, fences, and cattle. There were more than seven hundred dust storms during the 1930s, reaching a climax of severity around 1934 and

Dust storm in Colorado, 1930s. (Natural Resources Conservation Service).

1935. Eyewitness accounts of these dust storms were horrific: livestock were smothered to death, cleaning houses and clothes became a futile effort, and people grew sick and died from a host of respiratory ailments known collectively as "dust pneumonia." Visibility was so poor that one journalist quipped, "Lady Godiva could ride through the streets without even the horse seeing her." Soon the dust storms became so large that they blew soil from the plains all the way to the East Coast—sometimes depositing it on ships three hundred miles out in the Atlantic Ocean. The rest of the nation began to realize that this wasn't just Kansas's problem; it affected everyone.[24]

It was the Dust Bowl that finally gave Hugh Bennett a chance to do something about the nation's soil erosion crisis. In 1933, President Franklin D. Roosevelt created the Soil Erosion Service as a temporary agency to combat soil erosion—with Bennett at its helm. Bennett knew that the work he was doing was critically important, and he pleaded with Congress to make soil conservation a permanent national program. On March 20–21, 1935, Bennett testified before the House of Representatives, reciting fact after fact about the serious problem of soil erosion. Soon the congressmen began to notice that the sky outside was growing dark, although it was only midday, as another dust storm arrived from the Great Plains. As Bennett's biographer Wellington Brink described it, "Here were tons and tons of fertile soils—farms from Kansas and Colorado and New Mexico—swirling in from a 2,000-mile journey to tell the committee that this man Bennett was right, tragically right, urging them to accept his assurance that enormous folly was on the land and in the air and that something must be done immediately to stem it." On April 27, FDR signed the Soil Conservation Act of 1935, creating the Soil Conservation Service (SCS) in the Department of Agriculture—and Bennett's dream of having a permanent government agency devoted to soil conservation became a reality.[25]

The SCS's goal was to develop better farming methods that could stop soil erosion, increase agricultural yields, and keep land productive for generations to come. At first, soil conservation efforts focused on engineering methods, such as creating terraces in fields to catch rainwater and grading the sides of gullies to prevent further erosion. By the late 1930s, however, soil conservation relied increasingly on using vegetation to keep soil in place. The primary cause of soil erosion was bare soil on plowed fields and between row crops—something that never occurred in nature. As the ecologist Paul B. Sears put it, "Mother Earth is a staid and dignified old lady, no nudist by choice." If the soil could be kept covered by vegetation, fewer expensive engineering structures would be necessary. To ensure that their vegetative erosion control methods were effective, the SCS sought the advice of prominent plant ecologists like Frederic E. Clements (1874–1945). One important ecological concept that helped the SCS revegetate eroded areas was the observation that the natural regrowth of vegetation in a disturbed area followed an orderly sequence—a process known as *succession*. Soil conservationists discovered that they could accelerate succession by carefully selecting appropriate seeds and plants to stabilize gullies and heal eroded hillsides. Another practical ecological concept Clements developed was the use of *plant indicators*—remnants of native vegetation—to determine the types of plants best suited for a given area.[26]

Using these ecological concepts, the SCS developed simple but effective "vegetative methods of erosion control." Whenever possible, soil conservationists tried to keep the soil covered with some kind of plant. On the steepest slopes in humid regions like Ohio, this meant fencing cattle out of woodlots to allow natural forest vegetation to hold the soil in place. Moderate slopes were seeded in perennial grasses to serve as permanent pasture. Gullies were graded and planted in grass to serve as sod waterways. On the gentler slopes, old square fields were converted into strip fields that curved with the contours of the land. When it was necessary to plow the land to plant corn or other row crops, the rows were laid out to follow contours around hills instead of forming erosion channels by going up and down. A technique called *strip cropping*

Strip cropping in Coon Valley, Wisconsin, 1930s. (Natural Resources Conservation Service)

allowed some row crops to be grown on intermediate slopes. Fifty- to two-hundred-foot-wide contour strips of row crops were alternated with strips of tighter-rooted small grains like wheat, or with meadow (temporary pasture). These simple and practical methods significantly decreased erosion when used properly.[27]

The movement for soil conservation, headed by Hugh Bennett, was at its peak when Louis Bromfield returned to Ohio. But since he had been away from farming in the United States during the entire Depression and Dust Bowl period, soil erosion wasn't on Bromfield's radar when he bought his farm. While he wanted Malabar to be an "island of security" that he could come back to, Bromfield spent most of his first year of farm ownership in Hollywood with George Hawkins, writing scripts for the film version of *The Rains Came* and several other movies. His family stayed at Malabar in what Bromfield disparagingly called the "mail-order house" on the Beck farm until the Big House was completed. When they first moved in, this house had no electricity, only gas lights. Water was supplied from a hand pump in the cellar, and Ellen and Hope amused themselves by shooting bats in the attic with a BB gun. The day-to-day farm operations at Malabar were left to Bromfield's newly hired farm manager, Max Drake. Drake (1910–2001) had a degree in agriculture from the Ohio State University and had worked as a 4-H county agent in Medina for five years before he took the position at Malabar. He took

The mail-order house at Malabar, 1940s. (Malabar Farm archives, courtesy of Ohio Department of Natural Resources)

over farming operations in 1939, which mostly meant taking care of the handful of animals that came with the farm.[28]

In the spring of 1940, Drake realized that the fields at Malabar were going to need some serious soil conservation treatment, so he asked his friend Herschel Hecker for help. Hecker (1911–1991) also had an agricultural degree from OSU and had worked as a technician for the Soil Conservation Service since 1935, as part of SCS's extensive education and demonstration program. Hugh Bennett's preferred method for disseminating soil conservation practices was to create large soil conservation demonstration projects (25,000 to 30,000 acres) on a watershed basis, but individual farmers could also partner with SCS in single-farm demonstrations. An integral part of these demonstration projects was free labor provided by the Civilian Conservation Corps (CCC), a work-relief program implemented by President Roosevelt in 1933 to teach young men useful skills while conserving America's natural resources. Eight hundred of the nation's forty-five hundred total CCC camps were devoted to soil conservation, and Hecker alerted Max Drake to the fact that a new soil conservation CCC camp, Camp Rocky Fork, was opening near Mansfield in the spring of 1940. Thanks to Hecker's influence, one of the first places that the enrollees from the all–African American Rocky Fork camp worked was at Malabar. They laid out contour lines on the Malabar fields, filled and graded gullies, installed two diversion ditches on the hill behind the Big House to prevent more gullies from forming, and put in new fences to align with the land's contours.[29]

Louis Bromfield had no idea what was going on at Malabar, and when he returned from Hollywood in the summer of 1940 after having been gone for most of the previous year, he was shocked to see how much his farm had changed. "He kind of blew his stack," Drake later recalled. "I thought I was going to lose my job right then and there." After cooling down, Bromfield asked Drake to explain the purpose of the various soil conservation measures, and soon he was sold on the idea of soil conservation. Once he understood what was going on, Bromfield was pleased with the CCC's work. "They worked like beavers and they built fences that will be there, staunch and strong, as long as posts and wire hold out," he wrote later. Drake kept his job, and on November 13, 1940, he and Bromfield entered into a formal cooperative agreement for Malabar to be an official Soil Conservation Service demonstration farm. Hecker and other SCS technicians surveyed the farm's soils, topography, and current extent of erosion and plotted this data on a map. Next, they drew up a conservation plan for the farm, which specified the proper soil-conserving land use for each field. The SCS land use capability system, implemented in 1940, helped farmers develop cropping systems that were both ecologically and economically viable.[30]

Land use planning significantly changed the cropping pattern at Malabar Farm from what it had been historically. When Bromfield bought the property, 157 of the original 355 acres were in crops. When soil conservation technicians surveyed the land, they decided that only 93 acres could safely be used as cropland. On these crop fields, the former rotation of corn-oats-wheat-meadow was changed to a more soil-conserving rotation of row crop–winter grain–meadow–meadow. On some of the steeper fields, especially those northeast of the Big House, this rotation was used in strip cropping. The best use for another 137 acres was permanent pasture, and the conservation plan included applying lime and phosphate fertilizer to the pastures to boost the growth of soil-restoring grasses and legumes. Another eighty of the farm's acres were too steep to be safely used as pasture, and these were designated as woodland. Ollie Diller, a forester from the Ohio Agricultural Experiment Station, helped Bromfield create a forestry program, which included fencing the woodlots to keep cattle out so that young trees could grow. In return for this detailed planning and the free CCC labor, Bromfield and Drake agreed to follow the conservation plan for at least five years.[31]

Bromfield may have initially been skeptical of the new soil conservation practices, but when he saw them in action, he was quickly converted to the soil conservation cause. With soil conservation and fertilizer applications, the worn-out farm fields were yielding 30 to 40 percent more crops than they had just two or three years earlier. The "sickly, unnatural marsh" was converted into a watercress-filled pond,

Strip cropped fields at Malabar, 1940s. (box 69, folder 1496, Louis Bromfield Collection, Ohio State University, courtesy of Ohio Department of Natural Resources)

and the adjacent fields were reclaimed for vegetable production. Springs that had been nothing but mud puddles now flowed cool and clear, even in the driest part of the summer. Raw gullies and bare soil were healed over with a soft green blanket of grass and legumes, and soon the farm looked noticeably greener and more fertile than neighboring properties. "My travel, my experience—nothing I have ever done has given me nearly as much satisfaction as this bit of land and what we have been able to do with it," Bromfield said in 1941. "I take deep pleasure in going out every morning and seeing the miraculous changes which have happened, and which are happening, and which will go on happening until the end of our lives."[32]

# Chapter 2

# Crusading for Conservation

"After Pearl Harbor, I had to decide what as an American of middle years I could do best for country and the cause of freedom and security," Louis Bromfield told an audience of three hundred people in Cincinnati in February 1942. "I looked the whole thing over, and I decided that my job in this War was to devote my time and energies to fundamental things that will outlast any war and are eternal questions." Lounging on the platform, sticking his hands into the pockets of his blue flannel pin-striped suit, and not even looking at his notes, Bromfield broke all the formal rules of public speaking. But it didn't matter—once he started talking, everyone was spellbound. In big cities and small towns, to bankers, industrialists, garden clubs, conservation workers, and farmers, Bromfield made the dangers of soil erosion "immediate and terrifying." Each speech was slightly different, tailored to the audience and event, but the same message always came through: City people should care about conservation even more than farmers, because their business, industry, and food supply depended on a strong, stable agriculture.[1]

While Malabar remained Bromfield's "island of security" during the war, Bromfield was so excited about the results of the newly implemented soil conservation practices there—and so concerned about the continuing soil erosion and degradation on other farms—that his original dream of retreating into a rural hideout quickly disappeared. Within a year of his conversion to the soil conservation cause, Bromfield made connections with other crusading soil conservationists across the country and soon had a busy speaking schedule. In the first half of 1942 alone, Bromfield gave sixty speeches or radio broadcasts about conservation at various locations across the country. He never charged a speaking fee and insisted on paying for his own transportation, usually by train. As part of a small

Louis Bromfield (*center, seated*) and Hugh Bennett (*right, seated*) often shared lecture platforms at soil conservation rallies. This photograph was taken at Second Frontier Day in Ohio in 1947. (box 72, folder 1508, Louis Bromfield Collection, Ohio State University, courtesy of Ohio Department of Natural Resources)

but influential group of conservationists, Bromfield helped get conservation into the public's vocabulary during the war years as never before. Conservation was a patriotic duty, a necessity if the United States was to survive as a free and prosperous nation. And everyone—especially city dwellers—had a stake in the success of conservation.[2]

SCS chief Hugh Bennett, who often shared a platform with Louis Bromfield at soil conservation rallies and conferences, helped start a movement to educate the American public about the necessity of soil conservation. Bennett believed that the only way to solve the massive soil erosion problem in the United States was to inform people about the gravity of the situation and what could be done to remedy it. Under his leadership, the Soil Conservation Service published many well-written, engaging educational materials about soil erosion and conservation. To keep their material relevant to the general reader, SCS often contracted with writers and journalists from outside their agency to write these materials. One writer, a USDA employee named Russell Lord (1895–1964), was commissioned to write a long bulletin about soil erosion, its dire consequences, and what soil conservation could do to solve the problem. When Lord first set out on the assignment, he didn't "believe in erosion," but he had changed his mind by the time he finished the bulletin, *To Hold This Soil,* in 1938. "I believe in the menace of erosion now," he said. "I am afraid of it."[3]

In addition to this extensive publication program, Bennett wanted to establish a nonpolitical, nongovernmental organization to educate the public about conservation. Toward the end of 1939, at about the same time that Louis Bromfield returned from France and established Malabar Farm, Bennett traveled to Ohio and discussed this idea with two leading Ohio conservationists—Charles Holzer and Bryce Browning. Charles Holzer (1887–1956), a physician from Gallipolis, was well-known for his philanthropic involvement in his community—including constructing the town's first hospital—and was heavily involved in several conservation organizations, including the Ohio Valley Conservation and Flood Control Congress. Bryce Browning (1895–1984) was most famous for his leading role in the establishment of the Muskingum Watershed Conservancy District, a unique system of dams and reservoirs designed to stop floodwaters in the upstream tributaries of the Muskingum River before they could cause damage downstream. Along with Holzer and Browning, Bennett also consulted with Morris Llewellyn Cooke, an engineer from Philadelphia who had been involved in the field of conservation for many years. Cooke told Bennett that a taxi driver had recently asked him why there wasn't an organization to educate people about soil conservation. In January 1940, Russell Lord joined in the discussion and proposed that the society be called "Friends of the Land." With his journalistic assistance, these founding members typed up a foundational "Manifesto" explaining what Friends of the Land would be.[4]

"It is an old story, often repeated in the time of Man," the manifesto began. "We have hurt our land." Soil erosion was a major problem, resulting in loss of fertility and lowered agricultural production. "Soiled water depletes soil, exhausts underground and surface water supplies, raises flood levels, dispossesses shore and upland birds and animals from their accustomed haunts, chokes game-fish, diminishes shoreline seafood, clogs harbors, and stops with grit and boulders the purr of dynamos." The answer, the manifesto emphasized, did not lie in politics. Instead, it proposed "to bring quickly into action a non-profit association or society to support, increase and, to a greater degree, unify, all efforts of the conservation of soil, rain and all the living products, especially Man." The founders of Friends of the Land had remarkably ambitious goals from the very beginning: assembling and distributing information about conservation, forming regional groups, publishing a magazine, encouraging soil and water conservation research, getting conservation into school curriculums, organizing youth "in a moral equivalent of war against wastage of soil and water," cooperating with other organizations, and holding conferences. They wanted to educate everyone about conservation in every way possible. From the beginning, they planned this outreach to be international in scope, with hopes for a "World Conservation Congress" at some point in the future.[5]

Friends of the Land was officially inaugurated on March 22 and 23, 1940, when sixty men and women met at the Wardman Park Hotel in Washington, DC. They were from diverse backgrounds—including farmers, foresters, soil conservationists, educators, ecologists, editors, journalists, doctors, lawyers, and businessmen—but they had one goal: to establish an organization devoted to the conservation of soil and water. From the beginning, the society was quite informal, with much of the organizational work done in chairs on the hotel's "wide sun porch" overlooking a beautifully landscaped park. Even the planned sessions were kept spontaneous, with no "set speeches or long written papers" allowed and a ten-minute limit for each speaker. Practically everyone who attended had a chance to speak, from an Iowa farmer named J. J. Boatman to the well-known conservation advocates Stuart Chase and Jay N. "Ding" Darling. There was even one international representative—J. T. Detwiler, a biology professor from the University of Western Ontario.[6]

Louis Bromfield wasn't at the Friends of the Land organizational meeting; he was still preoccupied with writing movie scripts in Hollywood and wasn't fully committed to the conservation cause yet. Once he became convinced of the importance of soil conservation, however, it was inevitable that his path would cross that of Friends of the Land—and it happened when the new organization held its second annual meeting in Columbus, Ohio in July 1941. Along with other nationally known conservationists like Hugh Bennett, Paul B. Sears, and Stuart Chase, Louis Bromfield agreed to speak for the first day's meeting, at the Deshler Hotel on Friday, July 18—and everyone was blown away both by his skill as a public speaker and the passion of his message.[7] "It is quite true that my life has been somewhat divided between night clubs and manure piles," Bromfield began. He gave a short background of his childhood, his sojourn in France, and his decision to return to Ohio. Then he got into the meat of the story—the amazing success he had in restoring Malabar Farm. "The future of America is bound up in water and soil conservation," he concluded. "Americans face a job in these days which is far more important and more heroic than the job of the early pioneers. It is for each one of us to do his part, for each one of us is dependent upon the good earth in the end."[8]

Bromfield invited Friends of the Land to visit his farm and see his success for themselves, so the next morning they piled into a caravan of automobiles and headed toward Malabar. On the way, they stopped at the farm of Cosmos Blubaugh, an "amphitheater of plenty" among eroded hillsides. Blubaugh had worked for fifteen years to restore his farm's soils by adding every bit of organic matter he could lay his hands on, and the results were fantastic—high yields of corn, alfalfa, black raspberries, and apples, all planted on the contour in "semi-circular

*Above:* Cosmos Blubaugh farm, 1940s. (box 61, Friends of the Land Records Papers, courtesy of Ohio History Connection)

*Right:* Friends of the Land leaders on their July 19, 1941 visit to Malabar Farm. *From left:* Bryce C. Browning, Russell Lord, Morris L. Cooke, Louis Bromfield, Hugh H. Bennett, John D. Detwiler, and John F. Cunningham. (Malabar Farm archives, courtesy of Ohio Department of Natural Resources)

strips dyed various greens." The next stop on the tour was a lunch at Malabar Farm, where Bromfield's caterer had prepared a meal for 150 guests but had to rush into town to get more food, plates, and silverware for what turned out to be more than 500 people. After touring Malabar, about thirty cars continued on to visit nearby conservation landmarks such as the Muskingum Watershed Conservancy District, the Ohio State Agricultural Experiment Station at Wooster, and the SCS Hydrologic

Ollie Fink (*left*), Louis Bromfield (*center*), and other visitors to Malabar Farm, October 1941. (Malabar Farm archives, courtesy of Ohio Department of Natural Resources)

Station at Coshocton. Established in 1935, the Coshocton station was an experimental watershed that measured rainfall, infiltration, and runoff using lysimeters and other equipment. The tour continued on Sunday, when Friends of the Land visited southeast Ohio and saw the scars of clear-cut logging and strip-mining for coal—ruined land and depressed communities. Then they saw the progress being made in reforestation in the Hocking Valley, where hills that had been "as bald as an egg" twenty years earlier were covered with trees once again.[9]

After the Ohio conference and field trip in 1941, Bromfield became a devoted member of Friends of the Land. That same year, another Ohioan named Ollie E. Fink (1898–1970) joined the organization because of its emphasis on conservation education. Fink had a master's degree in education from OSU and was principal of the Hancock Junior High School in Zanesville in the late 1930s when he was asked to serve as an aide to George H. Maxwell, a well-known conservationist who helped plan the Muskingum Watershed Conservancy District. Soon Fink was converted to the cause of water conservation, and in 1937 he became very interested in a new movement for conservation education, spearheaded by the Soil Conservation Service, the Department of Interior's Office of Education, and the National Wildlife Federation. More than their Progressive Era counterparts, New Deal conservationists put a strong emphasis on education. This concurred with a broader shift in educational philosophy. In addition to their traditional role of teaching students the "three R's," most educational leaders in the 1930s believed that teachers should also educate their students about "current social and economic problems"—including conservation.[10]

In a democracy, conservation educators argued, education was the only means of solving social problems. An enlightened, informed citizenry would vote for leaders who made conservation a priority and would demand careful study by all disciplines of science before implementing conservation policies. Conservation was one of the most critical, if not *the* most critical, social and economic problems facing American civilization in the late 1930s—especially as the nation began to ramp up "defense" in preparation for entering World War II. Therefore, educating everyone about conservation, especially children in the public schools, was important, urgent, and patriotic. Instead of trying to add conservation as a separate class in an already crowded curriculum, conservation education advocates recommended that it be integrated into every class as an overarching philosophy to tie disparate subjects together. The key to making this approach succeed, they emphasized, was getting the teachers on board so that they could teach children about conservation with knowledge and passion.[11]

In 1938, the Zanesville public schools created a novel position for Fink: supervisor of conservation education for the entire school system. Fink developed what he called the "Zanesville Plan" to educate teachers about the importance of conservation so they could pass that knowledge on to their pupils. "As the land goes, business goes, property values go, and salaries go—down," Fink told the teachers, emphasizing that they had a personal stake in ensuring that conservation was practiced. Since few or no existing teaching resources were available, Fink wrote guides to help teachers integrate conservation into all their classes, from history to English, and provided them with reference materials from the Soil Conservation Service and other government agencies. He arranged field trips for teachers to see reforestation, soil conservation practices, and the flood control dams of the Muskingum Conservancy. Finally, he launched "a barrage of publicity" to "enlist the interest and approval of the public." Fink's conservation education work in Zanesville was so successful that in 1939, just a year later, he moved to Columbus to take a new position as the first-ever conservation curriculum supervisor for the entire state of Ohio.[12]

One creative method that Fink developed to give teachers hands-on training was the Conservation Laboratory, which he established at a group camp in the Tar Hollow State Forest in 1940. The goal of the Conservation Laboratory was to teach teachers "enough about man's

"Idle land does not contribute to the financial support of schools," Ollie Fink wrote in the caption of this photograph of him standing on gullied land next to a school in New Straitsville, Ohio, in 1941. (box 61, Friends of the Land Records Papers, courtesy of Ohio History Connection)

place in nature so that the importance of conservation will be clear to them" and so that they could pass this knowledge on to their pupils. At Tar Hollow, teachers stayed in rustic cabins, ate meals at a central lodge "flanked at either end by massive stone fireplaces," and spent their days out in the woods learning about nature and conservation. The students, divided into pairs, were assigned two-acre plots, containing both forest and field vegetation, to intensively study during the five- or six-week course. Students studied and mapped the "geological and soil conditions, topography, and animal and plant life" in their plots; hiked a three-mile nature trail; and contributed rock, soil, plant, and animal specimens to an on-site museum. The atmosphere was one of enthusiasm and discovery, and the small percentage of Ohio teachers who participated in the program went home committed to the conservation cause. "The list of never-to-be forgotten experiences is endless," one biology teacher wrote after attending. "Tar Hollow was and is a fine education!"[13]

In addition to his conservation education work, Fink became very interested in another hot topic in conservation—the relation of soil to health. "Poor soil makes poor people," warned Hugh Bennett, Louis Bromfield, and many other leading conservationists. Eroded, depleted soils, the reasoning ran, didn't have the mineral fertility to grow healthy plants—and so the people who farmed these soils and ate food grown on them were poor and sickly. "Soil debility soon removes stiffening lime from the national backbone, lowers the beat and vigor of the national bloodstream, and leads to a devitalized society," Friends of the Land wrote in their manifesto.[14] Fink had run into this theory while doing the research to write his conservation teaching guides, but he didn't devote much time to the topic until he met Dr. Jonathan Forman in January 1940.

Forman (1887–1974) earned his MD from the Starling Medical College in Columbus, studied for a year at the Harvard Medical School, and entered private medical practice in Columbus in 1920. His main interests were the study of allergy and medical history, and in 1936 he became editor of the *Ohio State Medical Journal,* which under his editorship grew to be "one of the top publications of the state medical societies." Forman was a strong opponent of socialized medicine and was publicly debating the topic with Kingsley Roberts, a New York advocate for medical cooperatives, when Fink first met him. When Forman argued that people would be healthier if they ate a nutritious diet, Fink jumped up from the audience and asked, "Dr. Forman, how can you expect to maintain the health of people from deficient foods produced on eroded and depleted soils?" After the meeting concluded, the two men continued their discussion and formed a friendship that took them both farther in the conservation movement "than either would have gone alone." Together, Fink and Forman joined Friends of the Land in 1941, bringing this emphasis on the connection between soil and health with them.[15]

As Fink and Forman discovered, the 1930s was an exciting time in nutrition research. Just a few years earlier, scientists had discovered the importance of vitamins and developed a list of vitamin-containing "protective foods" essential for health, including green leafy and yellow-orange fruits and vegetables, whole grains, whole milk, and animal organs like liver. The dentist and "nutritional anthropologist" Weston A. Price traveled around the world and concluded that people eating native diets of whole, natural foods had strong, healthy teeth, while those eating the vitamin-deficient "white man's diet" of white flour and sugar suffered greatly from tooth decay and general ill health. British nutritionist Sir Robert McCarrison conducted research in India demonstrating that rats fed whole wheat, vegetables, and milk were much healthier than those fed a deficient "Western" diet.[16]

The next major discovery in nutrition was that mineral elements were just as essential as vitamins. Animals and humans get most of their minerals from plants, and by 1940 scientists had made the intriguing discovery that plants grown on different soils had different mineral contents. Several previously puzzling regional animal and human diseases were found to be caused by deficiencies of certain mineral elements—iodine, selenium, and cobalt—in local soils and plants. This made many people, including University of Missouri soil scientist William A. Albrecht, wonder whether other common diseases might be caused by nutrient-deficient soils. Albrecht (1888–1974) looked at draft rejection and dental health statistics from World War II showing that young men from some regions of the United States, especially the Great Plains, were healthier than those from other regions, like the Deep South. Since the rich, black soils underlying the prairies of the Great Plains were more fertile than the red, leached soils of the South, Albrecht hypothesized that the food grown on them contained more minerals and therefore grew healthier plants, animals, and people.[17]

To test this hypothesis, Albrecht and his graduate students conducted several "biological assays of soil fertility" between 1939 and 1941. In seeming confirmation of Albrecht's soil-health hypothesis, these studies found that lambs and rabbits eating lespedeza hay grown on "better" Missouri soil types were bigger and healthier than those eating forage grown on "poor" soils. Fertilization improved forage quality and animal health on all soil types. But these results were criticized by other researchers, who pointed out that Albrecht didn't analyze the mineral contents of the soils or forages, the forage grown on unfertilized soils was mostly grass and weeds instead of lespedeza, and he used too few animals to tell whether the results were significant. The most valuable thing Albrecht's experiments did was get other researchers interested in the soil-nutrition relationship. In 1939, the USDA established the Plant, Soil, and Nutrition Laboratory on the campus of Cornell University, complete with greenhouses and state-of-the-art equipment for analyzing soil, plant, and animal samples. Because research on soil-health relationships was just getting under way when Ollie Fink first met Jonathan Forman in 1940, the USDA emphasized that "the data are wholly insufficient for defining such relationships in definite practical terms."[18]

Since the relationship between soil and health was such a hot topic, Fink and Forman decided to end the 1942 Conservation Laboratory session at Tar Hollow with a two-day conference on conservation, soil, nutrition, and human health. A hundred and fifty people assembled to hear William A. Albrecht, USDA researcher Kenneth C. Beeson, ecologist Paul B. Sears, and others speak on the theme "Our Health Depends on the Soil." The conference was so popular that it became an annual event, held as a wrap-up for the Conservation Laboratory in 1943 and 1944. In 1945, Friends of the Land officially took over the nutrition conferences, and they were held under the McGuffey elms on Ohio University's campus in Athens from 1945 to 1949. From 1950 to 1955, the soil and health conferences were held in Chicago in association with the College of Medicine at the University of Illinois. Conference topics over the years ranged from various aspects of soil fertility, plant health, and human nutrition to other conservation concerns, such as water conservation, population growth, and the dangers of pesticides. By the late 1940s, more than two hundred people were attending the

Leaders of Friends of the Land at one of the early nutrition conferences, circa 1942. *Top row, from left:* Jonathan Forman, Russell Lord, Charles Holzer, Walter Frye. *Middle row, from left:* Arthur Harper, Ollie E. Fink, C. Langdon White, Bryce Browning. *Bottom row, from left:* Kenneth C. Beeson, Paul B. Sears, D. R. Dodd, Frank Crow, John Detweiler, Wellington Brink. (box 61, Friends of the Land Records Papers, courtesy of Ohio History Connection)

conferences each year, with an all-time high of four hundred in 1947. In 1948 and 1949, Friends of the Land even published the complete conference proceedings in book format.[19]

Meanwhile, Louis Bromfield and Friends of the Land continued to promote soil conservation all across the United States, even as the nation geared up for all-out war production. Friends of the Land held their third annual meeting in St. Louis, Missouri, on July 23–25, 1942. After a stirring talk by Hugh Bennett about how soil conservation was the only way to produce enough food to feed the Allies and American troops overseas, the group headed off for an auto tour of Callaway County, Missouri, arranged by William A. Albrecht and other University of Missouri faculty. Each location was marked by a numbered plaque that corresponded to a printed program, and a loudspeaker car enabled

Caravan of automobiles at a 1944 Friends of the Land field trip in Alabama, including loud-speaker car at front right. (box 70, folder 1498, Louis Bromfield Collection, Ohio State University, courtesy of Ohio Department of Natural Resources)

the large group to hear what was going on at each stop as they looked at examples of soil erosion, restoration, and grass farming. That October, the group members met again in Memphis, Tennessee, where they listened to another round of stirring speeches and rode in an uncomfortable prisoner transport vehicle called "the Hoodlum" to see soil conservation in progress at the 4500-acre Shelby County penal farm. Thirty of the group, including Louis Bromfield and a "Latin American delegation," went on from there to tour the agricultural demonstrations of the Tennessee Valley Authority (TVA) and see what one of the New Deal's largest conservation agencies was doing to help farmers.[20]

The TVA was established on May 18, 1933 for the planned development of the Tennessee River Basin, encompassing parts of several states. While the best-known and most expensive part of the TVA was the construction of huge dams to control the Tennessee River and its tributaries and to provide hydroelectric power, the original program had many social elements as well, including an emphasis on agricultural improvement and soil conservation. In keeping with the conservation trends of the time, the TVA encouraged its farmers to practice soil conservation on their land, but with a twist. While the Soil Conservation Service was emphasizing contours, strip cropping, and terraces to control soil erosion, the TVA's soil conservation program emphasized growing grass and legumes to hold soils in place—with the aid of concentrated phosphate fertilizers. Near Sheffield, Alabama, was a defunct nitrogen-fixing plant, originally commissioned during World War I. The TVA took over the plant in 1934 and retooled it to produce concentrated superphosphate fertilizers from raw phosphate rock, using hydroelectric power from the Wilson Dam. To encourage farmers to use these phosphate fertilizers, TVA provided them free of charge to farmers and communities who agreed to participate in a demonstration program, on the condition that they plant leguminous crops to fix nitrogen. Neighboring farmers took notice, applied phosphate fertilizer themselves, and increased the region's agricultural productivity. Soon they were able to paint their houses, rebuild fences, dig ponds, and purchase electric appliances to use the cheap hydroelectric power they were getting from the dams—resulting, TVA boosters claimed, in prosperous, happy rural communities.[21]

At least, that was how it was supposed to work, though the reality wasn't always as perfect as the promotional literature made it sound. Many members of Friends of the Land were disturbed by the racism they encountered in TVA territory; Louis Bromfield drank out of a "colored" drinking fountain at Norris Dam to protest racial segregation. But the community that Friends of the Land was scheduled to visit on their TVA tour—Wheat, Tennessee—was described as a shining example of what an area agricultural demonstration could accomplish. More than half of the community's residents, on forty-three farms totaling 6800 acres, had improved their land and nearly doubled the number of livestock it could support. Cooperation in agriculture led to cooperation in other areas. By 1942, the community spirit in Wheat was so strong that the town had built a library with five hundred books, beautified its two churches, started a fair, served hot lunches at the school, learned to can and preserve food for the winter, and started holding weekly recreational meetings featuring music and folk dancing. Community members planned to build a cannery, locker refrigeration plant, and potato storage building in 1943. Louis Bromfield, Russell Lord, and other members of Friends of the Land were looking forward to visiting Wheat, but they were told that "to save transport" two representatives from the community would come and speak to them in the town of Norris instead. Only one person, Mrs. W. E. Gallaher, actually showed up—and everyone was shocked when, gallantly holding back her tears, she explained why.[22]

Wheat, its residents had just discovered, was being taken over by the federal government for some secret war project—along with a total of 59,000 acres on which over three thousand people lived. Residents didn't realize what was happening until they came home one day and found notices on their doors, informing them that the federal government was taking over their land for something called the "Kingston Demolition Range" and that they had only a few weeks to pack up and leave. Bewildered and angry, residents of Wheat and other communities in the condemned area were forced to move at a very difficult time of year. The government didn't pay enough for them to purchase farms of equal value to those they were leaving, and a shortage of trucks, gas, and tires made it very difficult to move their livestock and crops that they were still harvesting in October. For many, the most traumatic part was being forced to leave their ancestral farms; others had been displaced previously when the TVA built Norris Dam or when the Great Smoky Mountains National Park was established. But the TVA had always been careful to give people plenty of time to move and had scheduled evictions to avoid displacing people in the middle of the growing season; such considerations were completely lacking in this unprecedented military land grab.[23]

It was 1945 before anybody learned what the "Kingston Demolition Range" or "Clinton Engineering Works" really was—part of the top-secret Manhattan Project to make the atomic bomb that was dropped on Hiroshima. Oak Ridge, as the site was eventually known, was where uranium was enriched and then shipped off to Los Alamos, New Mexico, for the bomb's final assembly. Ironically, this remote site in Tennessee was selected for the Manhattan Project *because* of the TVA. Enriching uranium uses a tremendous amount of electricity, and the TVA's hydroelectric power plants produced some of the cheapest and most abundant electricity in the nation.[24] The paradox of the Tennessee Valley Authority was that each of its components could be used either for peace or for war. The phosphate furnaces could improve soil and grow alfalfa—or they could produce the raw materials for weapons designed to destroy homes and lives in Germany and Japan. Cheap hydroelectric power could raise living standards through rural electrification—or it could be used to power the factories that built the deadliest weapon in all of human history.

Friends of the Land toured Norris Dam in 1942. Constructed in 1936, Norris was the first major dam built by the TVA. (Photo by author)

With everything in the United States being turned to war, even conservation demonstrations, the future of Friends of the Land was in jeopardy. Who would be interested in soil conservation when natural security was at stake? "We picked a bad time to start," Morris Cooke wryly observed. And yet, soil conservation had to do with national security, too. What good would it do to defeat Hitler over in Europe if America's farms and fields were devastated in the process because of bad farming practices? The leaders of Friends of the Land believed conservation was just as important in war as in peace—maybe even more important. That was why Louis Bromfield devoted a large percentage of his time during the war to traveling and preaching for the conservation cause. And that was why Holzer gave generous donations to keep the organization afloat. "I'm not accustomed to failure," he said. "This is a mighty sick baby but there's life in it yet.[25]

The TVA visit was the last major field trip that Friends of the Land held for several years; soon thereafter, wartime travel restrictions confined them to smaller, regional gatherings. For the next few years, the organization's major outreach activity was its magazine, *The Land,* edited by Russell Lord. Despite the organization's precarious financial status, Lord believed a quality magazine would generate its own circulation, and so he arranged to have it printed by Waverly Press with a format similar to that of *Harper's* magazine. Published quarterly in green-covered 6½-by-10–inch issues of 100 to 150 pages each and illustrated with woodcuts by Kate Lord, Russell's wife, *The Land* was a unique publication from the very beginning. Lord included nostalgic articles about country life, updates on current conservation issues, proceedings of the society's meetings, forums debating controversial issues in conservation, a Foreign Correspondence section edited by assistant SCS chief Walter C. Lowdermilk, excerpts from both fiction and nonfiction books, and even poetry. The magazine's initial issues were reviewed favorably by leading editors and conservationists alike. Lord's favorite comment was from journalist and conservationist E. B. White, who wrote, "A magazine that is green inside and out is on the right track."[26]

Though the leaders of Friends of the Land enjoyed reading *The Land,* they were concerned that the society didn't really have enough money to publish such an expensive magazine. Bryce Browning and Louis Bromfield suggested putting out a smaller, less expensive periodical for the duration of the war, until the organization was on firmer financial footing and had a larger membership. "The idea of continuing to operate and to develop new debts without there being any assurance of a payment is just naturally 'rotten business,'" Browning explained. But the Lords had already developed an emotional attachment to their magazine and were "stunned" when they heard that the organization was considering suspending its publication. "It was partly as if our child and only living issue, The Land, were being smothered," Russell Lord wrote. Finally, a compromise was reached. The Washington business office was closed and the national office was moved to Bryce Browning's office in New Philadelphia, Ohio; and while publication of *The Land* continued, the second volume was thinner and set with smaller type to economize on paper. When more funds came in and Friends of the Land could pay their publishing bills, the length of the journal increased from a tightly packed 356 pages in 1942 to a rather overwhelming 636 in 1948.[27]

Meanwhile, the crusading efforts of Bromfield and other leaders of Friends of the Land started to pay off. Despite the war, or perhaps because of it, the membership of Friends of the Land continued to grow. The organization got a huge boost when Louis Bromfield wrote an article about it for *Reader's Digest* in 1944. Friends of the Land, Bromfield explained, was formed by "a group of unselfish citizens, alarmed by the terrible waste of our natural resources," who were aware that the only "sound method" to carry out reforms in a democracy was to educate the public. This article contained one of Bromfield's most concise and powerful summaries of the menace of soil erosion and urged readers to join Friends of the Land in their fight "to spread knowledge and information on this great national problem." Many people who read this article sent queries to the national office, now located in Columbus, asking how they could join Friends of the Land. To handle the increased correspondence and paperwork, Friends of the Land hired Ollie Fink to

Friends of the Land national office in Columbus, circa 1944. (box 71, Friends of the Land Records Papers, courtesy of Ohio History Connection)

work full-time in the Columbus office as the organization's executive secretary, starting May 1, 1944. After leaving his position with the State of Ohio, Fink divided his time between secretarial duties and traveling around the country, urging people to join Friends of the Land. And it worked. From a mere 1500 members in 1943, the ranks of Friends of the Land swelled to 3700 in 1945, 5500 in 1946, and 8600 in 1948.[28]

By the mid-1940s, Friends of the Land was firmly on its feet as the leading nontechnical soil conservation organization in the United States. It was also one of the first "co-educational" conservation organizations, with women taking on important organizational roles during the war years. "The care of the earth for all the years to come is not exclusively a male pursuit," Russell Lord explained. "We found in wartime that we simply could not get meetings arranged properly unless we let women help; do most of the work, in fact. That meant letting them speak in meeting, within due reason; and thus we came up with the astounding discovery that this made the meetings better." Thanks to the dedicated efforts of both male and female conservationists, Friends of the Land managed to not only survive the war years but grow and thrive. "The war has not been able to stop the activities of the Friends of the Land," Bromfield announced triumphantly in his 1944 *Reader's Digest* article. "I doubt that anything will ever destroy this organization, for its principles are those of democracy and common welfare, based upon hard economic fact and upon the self-interest of every citizen."[29]

Friends of the Land owed much to Louis Bromfield's literary and monetary contributions, and Bromfield benefited from the opportunity to collaborate with the nation's conservation leaders. The organization's annual meetings, nutrition conferences, and magazine served as discussion forums for all sorts of conservation issues. After he became an active member of Friends of the Land, Louis Bromfield used this constant exchange of ideas to turn Malabar Farm into something much more than just a run-of-the-mill SCS demonstration farm. While his farm managers oversaw day-to-day operations, Bromfield began using his land as a proving ground for the new ideas discussed at the Friends of the Land meetings and in the pages of *The Land*—and Malabar soon became one of the most famous private demonstration farms in the United States.

*Chapter 3*

# A New Kind of Pioneer

As Louis Bromfield had envisioned, Malabar Farm produced an abundance of food during the war, and no one living on the farm had to worry about wartime rationing. But the farm was no retreat from the worries and activities of the surrounding world—soon, the world was beating a path to Malabar. One night, Bromfield, always fond of the extravagant, hosted a thousand people at a Malabar Carnival featuring casino games, fortune-telling, and dancing under a full moon to raise money for the British War Relief Society. On other days, Malabar served as the venue for numerous local events, including the Richland County Fish and Game Protective Association's annual corn roast, a barbecue for the Nesmith Sportsmen's League, a spaghetti dinner for newspaper reporters from the *Mansfield News-Journal,* and a piano recital featuring the middle Bromfield daughter, Hope, playing selections from Beethoven.[1] Many people came to see Bromfield just because he was a famous author; others hoped to catch a glimpse of some of his Hollywood friends who might be visiting. And as time went on, more and more people came to see the soil conservation work that was progressing at Malabar Farm.

One of the first conservation-minded people to visit Malabar was a "smallish, graying man with very bright blue eyes" named Edward H. Faulkner. When Faulkner boldly announced in 1941 that he was developing a "new theory of cultivation which did away" with the moldboard plow, Bromfield thought he was crazy. The traditional moldboard plow, which completely buried all surface vegetation and crop residues to a depth of eight to ten inches to prepare the soil for planting, was the cornerstone of American agriculture. Thomas Jefferson improved the moldboard's design in 1784, Charles Newfold patented the idea in 1796, and in 1837 an enterprising blacksmith named John Deere began manufacturing steel

Moldboard plows, even those pulled by horses, left the soil bare of vegetation and susceptible to erosion. (Natural Resources Conservation Service)

moldboards in Illinois. "Sodbusting" moldboard plows enabled pioneers to move west of the Mississippi and convert millions of acres of prairies into cropland. The moldboard plow was more than just a tool; it was a symbol of American independence, entrepreneurship, and innovation. To suggest its retirement from agriculture was as heretical as "criticizing the design of the American flag"—or proposing "that the industrial world do away with the locomotive or the blast furnace."[2]

But, as Faulkner continued to talk, Bromfield started to realize that maybe he was onto something. Faulkner (1886–1964) had been born on a small farm in hilly Whitley County, Kentucky, near the Tennessee border. His father was a successful farmer who restored worn-out land and grew high-quality vegetables for the local market using a combination of chemical fertilizers, manure, and crop rotations. In general, however, the quality of agriculture in his hometown was poor, so Faulkner decided to study agriculture at the University of Kentucky. He worked as an extension agent in Kentucky for four years and then moved to Ohio, where he continued his career as a county agent, agricultural teacher, and soil investigator. By 1930, Faulkner had retired to a small piece of ground near Elyria, Ohio, where he conducted tillage experiments in his backyard and a rented field. Based on his experiments, Faulkner concluded that moldboard plowing was detrimental to soil health because it buried crop residues so deeply that they couldn't decompose properly or release nutrients into the plant rooting zone. Without the continual renewal of organic matter in the upper layers of the soil, any residual humus would rapidly be oxidized, leaving the soil bare, infertile, and vulnerable to erosion. The moldboard plow, Faulkner claimed, was the main culprit in the soil erosion disaster that was sweeping the United States.[3]

The key to rebuilding soil, Faulkner believed, was to mix organic residues into the soil instead of burying them with a moldboard. Faulkner predicted that if this kind of "trash farming" was widely adopted, soil organic matter would increase so much that fertilizers would become unnecessary and pests and weeds would eventually disappear. With assistance and encouragement from Louis Bromfield, Ollie Fink, and the ecologist Paul B. Sears, Faulkner decided to publish his ideas in a book. Several large publishing houses rejected the manuscript, but finally Savoie Lottinville, at the University of Oklahoma Press, "had the courage to publish it." To everyone's surprise, *Plowman's Folly* became a national bestseller. Soon everyone was talking about it, from farmers to actresses, and the University of Oklahoma Press could not print enough copies to satisfy the demand. By 1948, over a million copies of *Plowman's Folly* had been sold, including a pocket edition for soldiers. Excited by this success, Faulkner went on to write two more agricultural books: *A Second Look* in 1947 and *Soil Development* in 1952.[4]

Faulkner's ideas about tillage were highly controversial. Many agronomists criticized his book, calling it "Faulkner's Folly." William A. Albrecht believed "Faulkner's indictment of the plow will not stand against the facts of science nor the judgment of experienced farmers." Others, including Hugh Bennett, thought that the book was very timely. "Like fine wine, fast automobiles and cussin' words, the old moldboard or turning plow is something that serves us best when used with moderation," Bennett explained. "But it takes a strong man to resist the temptations of excess; and an experienced man to see the wisdom of discriminate use. For the past two hundred years, more or less, we in the United States have been on something pretty closely akin to a moldboard plow jag." Many people thought Faulkner's criticism of the plow was valid but believed some of the other theories he included in his books—such as soil minerals being "inexhaustible" if the soil contained organic matter, or that capillary water rising from the subsoil could supply all the water needs of plants regardless of rainfall—were a little extreme and not backed up by scientific evidence. Nobody, Faulkner noted, criticized his claim that soils rich in organic matter were superior for optimum crop production and soil conservation.[5]

As the debate over *Plowman's Folly* heightened, it turned out that Faulkner was not the first American to try non-inversion plowing. Down in Hall County, Georgia, a farmer named Mack Gowder had been using a homemade non-turning plow, which he called a "bull-tongue scooter," for thirty years before Faulkner's book was published. Christopher M. Gallup, from Stonington, Connecticut, had been using a spring-tooth harrow for primary tillage for ten years. During a 1944 speaking tour in Alabama, Louis Bromfield, Hugh Bennett, and other Friends of the Land members visited the farm of David Yarbrough in Autauga County, where they saw a cornfield that had just been fitted and planted with a "bull-tongued, non-turning plow." Conservation tillage was even more popular in the Great Plains, where soil conservationists had been telling farmers since the early 1930s that one of the best ways to halt wind erosion was to leave crop residues on the soil surface. An Oklahoma farmer

*Right:* Mack Gowder's "bull-tongue plow," 1940. (SCS photo, box 61, Friends of the Land Records Papers, courtesy of Ohio History Connection)

*Below:* Louis Bromfield (*left*), R. Y. Bailey (*center*), and David Yarbrough (*right*) examine a stubble-mulch field in Alabama, 1944. (box 70, folder 1498, Louis Bromfield Collection, Ohio State University, courtesy of Ohio Department of Natural Resources)

Louis Bromfield using the Seaman rotary tiller. (box 69, folder 1494, Louis Bromfield Collection, Ohio State University, courtesy of Ohio Department of Natural Resources).

named Fred Hoeme started designing a chisel plow—an extra-sturdy harrow that could break up hard ground but did not invert the soil—in 1933. After manufacturing about two thousand chisel plows, Hoeme sold the production rights to W. T. Graham in 1937, who manufactured the "Graham-Hoeme" plows in his Amarillo, Texas factory. By the early 1950s, an estimated half of the farmers in the Great Plains were using chisel plows to control wind erosion on their land.[6]

By the mid-1940s, Louis Bromfield had become a pioneer in conservation tillage methods. While he didn't give up using the moldboard plow altogether, he discovered that "trash farming"—leaving crop residues on the surface—was one of the best ways to plant alfalfa on his hilly Ohio soils. At first, Bromfield used makeshift implements like a cultivator and a disk to "rip up the soil without turning it over," which "left the field with a surface mulch of rotting roots and stems which acted all through the winter as a kind of sponge" and grew "an extraordinary stand" of alfalfa in the spring. But with his knack for making friends in high places, Bromfield soon got on a first-name basis with many of the farm machinery manufacturers in the United States. They would visit Malabar, see the number of people who were visiting or reading Bromfield's books, and then loan Bromfield some of their best farm machinery for him to demonstrate at Malabar. It was a mutually beneficial relationship—the manufacturers got lots of free publicity, and Bromfield got lots of free machinery. His two favorite conservation tillage implements were the Graham-Hoeme chisel plow and the Seaman rotary tiller, manufactured by the Seaman-Andwell Corporation of Milwaukee, Wisconsin.[7]

As time went on, Bromfield experimented with different systems of conservation tillage and finally settled on a method that he called "sheet composting" or "deep tillage." Rather than merely scratching the surface of the soil with a disk like Faulkner recommended, the idea of sheet composting was to plow "as deep and roughly as possible," mixing in crop residues to a depth of nine or ten inches with the Graham-Hoeme plow and the Seaman tiller. Bromfield emphasized deep, non-inversion tillage because he found that the gravel subsoil at Malabar was more fertile than the depleted topsoil. He claimed that he could build topsoil out of subsoil at the rate of about an inch per year by using conservation tillage and planting fast-growing, deep-rooted grasses and legumes. Bromfield also discovered that if he used conservation tillage, many engineering soil conservation practices became unnecessary. As soil organic matter increased, he could widen his contour strips from 75 to 150 feet without increasing erosion. One of his neighbors found that two gullies on his property stopped eroding once Bromfield checked runoff uphill with his trash farming. And the big diversion ditch behind the Big House, which the CCC had built under Herschel Hecker's supervision, remained completely dry and was nicknamed "Hecker's Folly." Louis Bromfield foresaw, rightly, that conservation tillage would play a major role in future American soil conservation.[8]

Even the best tillage implements would not work without a tractor to pull them, and one of the secrets to Bromfield's success in conservation tillage was the Ferguson tractors he used. Bromfield didn't have much experience with tractors before he purchased Malabar Farm; back when he was a young man working on his grandfather's farm, all the work was done by horses. Huge steam traction engines were used in the giant wheat fields of the Great Plains, but they were too big, expensive, and inflexible for smaller Ohio farms. It wasn't until Henry Ford released the Fordson tractor in 1917 that the average midwestern farmer could even think about getting a tractor. Other manufacturers quickly followed suit, and by 1940 most farmers who could afford them—and many who could not—had purchased tractors. The Herring family that Bromfield bought the farm from in 1939 apparently did not own a tractor, however, because all Bromfield got with the farm was a single team of draft horses.[9]

Despite their popularity, all of the early tractors had a major flaw—they often tipped over during plowing. The first tractors used modified horse-drawn plows that were towed like trailers. The problem with this was that a tractor was much more powerful than a horse, so if the plow suddenly hit a rock or other obstruction, the force would cause the tractor to rear up and fall over backward, often killing the driver unless he quickly slammed down the clutch pedal. An Irishman named Harry Ferguson (1884–1960) finally overcame this design flaw in the 1930s by perfecting a novel plow attachment system—the *three-point hitch.* When a plow attached using the three-point Ferguson System hit an obstruction, the extra force was transferred to the front end of the tractor, allowing the rear wheels to spin harmlessly instead of tipping the tractor over.[10]

The only drawback of the Ferguson System was that to work properly, it required a complete redesign of the tractor, and none of the leading tractor companies were willing to take the risk of adopting a novel design that would not be interchangeable with any of their older implements. After a couple partnerships in England, Ferguson traveled to Detroit and made a famous "handshake agreement" with Henry Ford. Ford agreed to produce tractors with the Ferguson-designed implement attachment system at his Dearborn factory, with Ferguson in charge of engineering. With the Ford Company to back it and the inherent advantages of the Ferguson System, the new tractor was a success. More than 10,000 tractors were sold in the first year, and over 42,000 were produced in 1941. The United States' entry into World War II and the subsequent restrictions on manufacturing resulted in a huge drop in production in 1942, but after Ferguson personally demonstrated his tractor for President Franklin D. Roosevelt, the company had no trouble getting supplies for 1943.[11]

Louis Bromfield first met Harry Ferguson in 1943 and was fascinated both with Ferguson's tractors and a grandiose economic scheme called "The Plan" that Ferguson advocated with missionary zeal. Ferguson, whom Bromfield described as "a small man, lean and looking rather like a benevolent bird," was obsessed with his plan to "eliminate the power animal on every farm in the world." He believed that all of the world's problems stemmed from inflation, which was caused by inefficient production. And nowhere was production more inefficient than in agriculture, where big draft animals not only made farming slow and difficult but also required a huge amount of land to grow their food. Ferguson's proposal was, frankly, one of world domination—to replace every team of horses in the world with a Ferguson tractor, because no other tractor could do the job. Ferguson "had the ability of making you feel that what he was trying to do was the only thing in the world worth doing," one of his employees noted. And one of the people he persuaded to endorse his ideas was Louis Bromfield, who agreed to write an article about the Ferguson Plan for *Reader's Digest.*[12]

Once he actually started on the article, however, Bromfield found Ferguson quite difficult to work with. Ferguson wanted the whole article to be about his proposed economic system, while Bromfield wanted to include biographical information to give the story more human interest. The final published article said little about the Plan and read more like an advertisement for the Ford-Ferguson tractor, prompting the International

Louis Bromfield on a Ferguson tractor. (Malabar Farm archives, courtesy of Ohio Department of Natural Resources)

Harvester Company to send a long letter to the editor complaining that Bromfield had focused on just one brand of tractor. "I have been accused of favoring the product of one farm machinery company over all the others," Bromfield told the Economic Club of Detroit in 1945. "That is true and I did so and continue to do so because that company has shown vision in producing a whole system of machinery which is light, modern, practical, convenient and is designed for efficiency and with the needs of *all* farmers in view." Nevertheless, he didn't publish any more articles promoting Ferguson tractors, and Ferguson's suggestion of having Bromfield write a whole book about his Plan never materialized. The closest he got was writing a booklet for the Ferguson company in 1952, *The Wealth of the Soil,* which emphasized that good soil-conserving farming was profitable but was pretty toned down from Ferguson's grandiose economic ideas of the early 1940s.[13]

The most important thing to come from Bromfield's relationship with Harry Ferguson was a deal that Ferguson would loan four tractors to Malabar for demonstration purposes and would exchange these tractors "every two years in order to have these units up-to-date and in first-class condition." No contract between Bromfield and Ferguson

Ferguson tractor with chisel plow. (Malabar Farm archives, courtesy of Ohio Department of Natural Resources)

survives to give the details of this arrangement; likely it was a "handshake agreement" like the more famous one between Ferguson and Ford. The first Ferguson tractors arrived at Malabar in the spring of 1944, and by September Bromfield had already broken two plows. The agreement between Bromfield and Harry Ferguson lasted for the rest of Bromfield's life, and probably every model of tractor that Ferguson produced between 1944 and 1954 ended up at Malabar for at least a short time. Harry Ferguson split with Ford in 1947 to form his own tractor company, Harry Ferguson, Ltd., which merged with the Toronto-based Massey-Harris company in 1953 to create Massey-Harris-Ferguson (the name was shortened to Massey-Ferguson in 1958), but through it all there were always brand-new Ferguson tractors at Malabar Farm.[14]

With a combination of new knowledge about the importance of soil organic matter, new conservation tillage implements, and the new Ferguson tractor to pull them, Bromfield achieved astounding success in restoring his worn-out fields to productivity. His favorite story to tell about soil restoration at Malabar was "the miracle of the Bailey Hills." Known as "the poorest farm in Pleasant Valley," the "Bailey" farm was in such bad shape that Max Drake cried when he heard that Bromfield had purchased it in April 1942 from Laura Niman. Bromfield bought the farm mostly for the historic brick farmhouse, a huge spring that flowed out of the sandstone cliff, and the wonderful view from the highest hill. But that hill was known locally as "Poverty Nob" because the soil was so depleted that it grew only "wire grass, poverty grass, and broom sedge," which provided some pasturage early in the year but turned dead and brown during the summer. The only green spot was on the

"Poverty Nob" (Mount Jeez) in summer before restoration. Note that vegetation on the top of the hill is greener (darker) than the rest of the hill. (Malabar Farm archives, courtesy of Ohio Department of Natural Resources)

top of the tallest hill, where grazing sheep had "gleaned what little fertility remained and carried it to the hilltop where they deposited it as manure." In summer, Bromfield wrote, the hill "resembled a brown, sterile mountain capped by emerald-green snow."[15]

Restoring the hills was going to be a challenge, but Bromfield delighted in doing what many farmers believed impossible. First, he "mined" several decades worth of accumulated manure and old straw from the barn and spread it on the fields, along with some lime to raise the soil pH enough to plant legumes. Next, he used conservation tillage to break up the old, acid-loving vegetation and mix it into the soil, then planted alfalfa, brome grass, and ladino clover. When a huge crop of weeds emerged along with the forages, he just mowed them down and used the mulch to add even more organic matter. The soil, he discovered, wasn't really worn-out at all—it was just low on organic matter, and conservation tillage was the quickest way to remedy that deficiency. A year after Bromfield started restoration work, the formerly barren hills were covered with a "deep emerald-green" blanket of grass and legumes "from top to bottom." After three or four years of conservation tillage, green manuring, and fertilization, some of the lower fields on the formerly worn-out farm yielded fifty bushels of wheat per acre—more than twice the Ohio average. And that was just one field—similar miracles of soil restoration were taking place all over Malabar Farm.[16]

Bromfield knew he was on to something important. As his yields increased and the scars of erosion caused by decades of bad farming healed over with grass and legumes, he became so excited that he just couldn't keep it to himself. In 1945, for the first time in his literary career, he wrote a nonfiction book. In *Pleasant Valley,* Bromfield told his readers in lyrical, descriptive prose how he had purchased Malabar, shared some of the rich history of the area, and described the phenomenal progress that he had made by working with natural processes to restore the land. And Malabar Farm, Bromfield told his readers, was only the beginning. All over the United States, thousands of similarly worn-out farms were just waiting for someone who understood soil conservation and the importance of working with nature to begin conservation work. "What we needed was a new kind of pioneer, not the sort which cut down the forests and burned off the prairies and raped the land, but pioneers who created new forests and healed and restored the richness of the country God had given us," Bromfield explained. He considered himself "one of that new race of pioneers" and urged his readers to join him in the critical and rewarding work of restoring the nation's farmland.[17]

The timing for the book's release couldn't have been better. The Americans who read Bromfield's books had just lived through the Great Depression, with the horrors of the Dust Bowl, the stock market crash, and people literally starving because they couldn't afford food. Then came World War II, which restored economic prosperity but also brought rationing of civilian food supplies (especially things like sugar and meat) to provide food for the military and the Allies. "Food Will Win the War and Write the Peace," proclaimed Secretary of Agriculture Claude R. Wickard, and farmers were encouraged to produce as much food as possible. A similar plea had been made during the First World War, but this time there was a crucial difference: farmers were told to increase production by conserving their soil. "Only by going 'all out' for conservation of all our natural resources—and foremost our soil—can we keep America virile and free," warned Hugh Bennett in 1942.[18] As the war was ending in 1945, with Allied victory already practically assured, one of the most important topics of conversation was what the postwar world would look like. Would there be another economic collapse in the 1950s and another depression? Or could the world finally work together to create a stable global society that would provide peace and prosperity to everyone for all time?

"The time to cast iron into a new mold is while it is molten," agricultural economist Howard R. Tolley argued in 1945. "Human institutions, habits, and values throughout the world are now in flux and will remain so until the reconversion from war to peace is ended. . . . As soon as the war is over—perhaps even sooner—they will begin to crystallize and harden into the pattern of the postwar world." The hope of Bennett, Bromfield, and other conservationists was that permanent agriculture, based on soil conservation and improved farming methods, would be the foundation of that new world pattern. Conservationists dreamed of an economically and ecologically stable farming system that included better recycling of organic waste materials, allowed farmers to make a living by using more efficient production methods, and lowered food costs for consumers at the same time. "I believe that one day our soil and our forests from one end of the country to the other will be well managed and our supplies of water will be abundant and clean," Bromfield wrote in *Pleasant Valley*. "I believe that there will be abundance for all as God and Nature intended . . . there will be no more floods . . . the abomination of great industrial cities will become a thing of the past. . . . The possibilities of the future are boundless."[19]

As soon as travel restrictions were lifted after the war, people started coming from all over the country to see Bromfield's restoration work for themselves. Once they arrived at Malabar, visitors were often surprised to find the world-famous author wearing a grungy plaid flannel shirt and "old corduroy trousers" and, as often as not, driving a tractor or pitching manure. Once they got over the initial shock, most visitors were impressed to see Bromfield actually doing farmwork. "You know that he is a dirt farmer with work on his mind," one visitor observed. When he lectured passionately on the importance of soil conservation, his listeners could see for themselves that he really practiced what he preached. "Louis Bromfield's 'Pleasant Valley' is all and more than he

Louis Bromfield (*second from right*) leading a group of military personnel across Switzer's Creek, early 1940s. (Malabar Farm archives, courtesy of Ohio Department of Natural Resources)

Louis Bromfield (*right*) talking to a group of visitors on Mount Jeez. (A/V box, from box 11, Friends of the Land Records Papers, courtesy of Ohio History Connection)

pictures it in his popular book by the same name," one reporter enthusiastically wrote in 1946 after visiting Malabar. "Surprising to those who may glorify Malabar from afar after reading a book which tells a good story, it is a FARM!"[20]

Unless they made special arrangements, Bromfield let visitors wander around the farm themselves during the week. On Sunday afternoons in the summer, he gave public lectures on whatever aspect of agriculture or conservation he was most interested in at the moment, then offered to take visitors on a "quick look around," which was really a strenuous several-hour tour of the entire farm. Bromfield designed his tour to weed out mere curiosity-seekers and celebrity-worshipers so only devoted farmers made it to the end—an overview of the farm from the highest hill, renamed Mount Jeez after it was restored to productivity. Most visitors, even fashionable ladies in high-heeled shoes, followed Bromfield around the farm on foot, dodging cow pies and dewy grass as they trekked across pastures. Newspaper reporters and other favored guests got to ride with Bromfield in his new Jeep, but they had to fight with his Boxer dogs for a seat. As they traveled across the farm, Bromfield told all his visitors the same story about how soil conservation had increased yields, turned brown hills green, started long-dry springs flowing again,

Bromfield giving a group of ladies (possibly a garden club) a tour of the gardens by the Big House. (Malabar Farm archives, courtesy of Ohio Department of Natural Resources)

and was supporting an ever-increasing number of ever-healthier plants and farm animals.[21]

Hundreds of groups—small and large, young and old, men and women, influential leaders and farmers—traveled from near and far to see Malabar. By 1947, anywhere between a hundred and a thousand visitors might tour Malabar on a typical summer Sunday afternoon. Bromfield's favorite visitors were the farmers, including a group of over five hundred Amish men and women from Holmes County, Ohio, who made special arrangements to tour on Thursday instead of Sunday. "I would receive the Amish—particularly the younger and more modern ones—any time for they are the best farmers in the world and have a better instinctive understanding of soil and livestock than any group I know," Bromfield said. He was especially impressed with the Amish women, who often read more farming publications than their husbands did and had in-depth knowledge of agriculture, biology, chemistry, economics, and soil health. "If all of us had possessed the Amish philosophy, the cost of living in this country would be about forty to fifty percent less, the food would be more nutritious by an equal per centage, taxes would be much lower and there would be no price supports nor any subsidized agriculture," Bromfield wrote in his journal after their visit.[22]

Most visitors—especially those who came in groups—brought a picnic lunch to eat either at the farm or in the nearby Mohican State Forest. Only a few carefully selected guests each day were invited to have lunch or dinner in the Big House with the Bromfield family, and they invariably recorded lunchtime as one of the most interesting experiences at the farm. The food was always excellent and abundant. One visitor recalled a menu of "large, light pancakes," "plump, tender cakes of sausage," tomatoes, watercress, lettuce, cheese, honey, maple syrup, and butter. Another group of visitors had "an omelet with chives, served on buttered toast, baked potatoes, bowls of mixed vegetable salad, three or four kinds of bread, pound bricks of butter, melons, milk." The lunchtime conversation could get quite lively, especially with the random assortment of people often at the table together. When one reporter visited, there were ten people at the lunch table, which was "much like any big-family gathering." The Bromfield family members teased a young French farm worker about his Brooklyn accent, expressed their distaste for football, and "chatted away about any number of subjects, not ignoring the guests and yet not insisting that they take an active part."[23]

Almost every visitor to the farm noted that Bromfield was always surrounded by his faithful Boxer dogs—Prince, Baby, Gina, Folly, and several others—whether he was in the house, mowing a field of hay, or driving around the farm in his Jeep. Some visitors—even those who loved dogs—were initially intimidated by the flood of canines that surrounded their cars when they pulled up in front of the Big House. Boxers, the nationally syndicated columnist Inez Robb noted, were "the droolingest dogs ever invented." Despite weighing about seventy pounds each, they tried to sit on the laps of guests who were seated at the table, begging for food scraps. The Boxers ruled the roost at Malabar; if dogs were occupying all of the chairs in the living room, guests weren't allowed to kick them off while Bromfield was in the room. The Boxers learned how to open car doors and close them once they got inside, and several visitors to Malabar found five or six dogs sitting in their cars when it was time to leave. One time, this almost ended in tragedy when Bromfield left two of the Boxers in his car, parked on a slope, while he ran inside to use the telephone. He came back out of the house just in time to see the car rolling down the hill into the pond and immediately dived into the water to open the car door and save the dogs. Only once, according to

Inez Robb, was there a "rift between Mr. B. and his beloved dogs"—when a man came to the door and asked the Boxers, instead of Bromfield, to autograph his copies of Bromfield's books.[24]

In addition to the Boxers, Malabar Farm was full of "characters," both animal and human. When visitors rang the doorbell, they were often greeted by a gobbling tom turkey. Or they might see four goats calmly ruminating on the porch swing and come back to find them on the roof of their automobile; Bromfield finally found the goats a new home when he discovered them chewing on a pile of letters inside someone's car. There were bottle-fed lambs who thought they were dogs, a duck named Donald (who turned out to be female) who preferred for a long time to live at the Big House rather than with other ducks, Haile the Karakul ram, Blondy the pure black Angus bull, and Sylvester the Guernsey bull—plus numerous cows, cats, and other animals. Then there were the people who lived on the farm. Louis Bromfield, with his intense and charismatic personality, was of course the main attraction, but there were also his three daughters—Anne, Hope, and Ellen—and his quiet and gracious wife, Mary Bromfield. While Mary appreciated the beautiful landscape at Malabar and enjoyed entertaining her friends at the farm, she was overwhelmed by the sheer number of visitors who came, especially on Sundays. Her ideal Sunday would have been spent "lolling on the lawn and gazing at the landscape all of the afternoon" with a small number of invited guests; instead, her guests who came one Sunday didn't even get a chance to see the farm because of the over five hundred uninvited visitors who invaded the yard.[25]

Bromfield (*front left*), visitors, and bored-looking Boxers on the lawn in front of the Big House. (box 70, folder 1498, Louis Bromfield Collection, Ohio State University, courtesy of Ohio Department of Natural Resources)

Anne, Hope, and Ellen on a fence at Malabar. (Malabar Farm archives, courtesy of Ohio Department of Natural Resources)

In complete contrast to Mary was George Hawkins, a short, plump man who had lived with the Bromfield family as Louis's manager since 1929. Hawkins had an attitude of unmitigated disgust toward the entire farming venture and thought Bromfield should have stuck with writing novels instead of "humus, mucus, retch and vetch" for his farm books. He would shock visitors to Malabar by appearing outside in his bathing trunks and saying, "God will be with you in a moment"—referring to their seemingly religious devotion to Bromfield. Ellen said that it was Hawkins who named the highest hill on the farm "Mount Jeez," after Bromfield's weekly "Sermons on the Mount"; Bromfield's version of the story was that a young farm worker named it because of the breathtaking view. Even though he didn't believe in the conservation cause, Hawkins accompanied Bromfield on many of his speaking tours around the country. "We are

Mary, Ellen, Louis, and Boxers in a typically cluttered Big House. (Malabar Farm archives, courtesy of Ohio Department of Natural Resources)

doomed!" he concluded satirically after attending one of the Friends of the Land meetings. "I don't think there is much to do but to go and cut my throat. It is all over!" The meeting was a success, he reported, "but nobody seems to be doing anything about the farmer. He still goes on farming in the old way—greedily 'mining' the soil and topsoil (It must be the deepest nine inches in the world) is still rushing into the Gulf of Mexico. I am afraid there is nothing more to be done. We are doomed!" Despite his distaste for their soil conservation rhetoric, Friends of the Land enjoyed having Hawkins along at their meetings. "He has such a gay time having an awful time in the wide open spaces that we wouldn't be without his company on our tours," Russell Lord explained.[26]

Most of the people who visited Malabar were profoundly affected by the experience. Some actually implemented soil conservation programs on farms of their own. Others were just excited to find that Bromfield's *Pleasant Valley* account was true. But the most observant visitors noticed that a few things didn't line up with what Bromfield had written in his book. "It looks convincing, but where are your potatoes?" asked one reporter who toured the farm in 1946. When Bromfield explained that they didn't grow any potatoes, because they could buy them cheaper, the reporter said, "Now the way I read Pleasant Valley, every farm should be self-sufficient and grow a little of everything the family needed." "Well," Bromfield replied, "I'm being persuaded that maybe that's not entirely right."[27]

George Hawkins and Ellen in front of the Big House. (box 69, folder 1497, Louis Bromfield Collection, Ohio State University, courtesy of Ohio Department of Natural Resources).

The truth was Malabar only followed Bromfield's original plan for self-sufficiency for a few years. During the war, the self-sufficiency aspect of Malabar allowed the workers to have ample supplies of meat, butter, honey, maple syrup, and other products that were expensive or completely unavailable at stores. Once prices went down after the war, however, Bromfield found that his original "general farm" idea was not economically viable. He could not grow corn and compete with farmers in northwest Ohio who had thousands of acres of perfectly flat land at their disposal. The two-hundred-tree apple orchard was too big to manage and too small to be profitable; a similar situation applied to the chicken, sheep, and hog operations. Instead of trying to produce everything, Bromfield decided to specialize in the crop that Malabar Farm was best suited for—grass.[28]

## Chapter 4

# The Golden Age of Grass

"If we had never heard of grass farming we should have become grass farmers simply through the evidence of our own senses," Louis Bromfield wrote in 1948.

> The advantages showed up in the deepening color of the soil, in the miraculously increasing yields, in the evidence of the farm's account books, in the flavor and tenderness of the meat, in the sleekness and shininess of the coats on our cattle, and the brightness of their eyes, even in the changing, ever-augmenting beauty of the landscape as grass and legumes healed over the old gullies, the poor spots in already poor fields, cleared the once muddy streams, brought life to dead or dying springs, and saved the rainfall which made the very trees more green and luxuriant in appearance.

After the return to a peacetime economy put an end to Bromfield's dreams of self-sufficiency, he decided to specialize in just one main crop at Malabar—grass. While grass had always been part of Bromfield's soil conservation program, it had been only one component of his original "Plan." But by 1950, Bromfield found that grass farming was the perfect fit for Malabar. If he had started out in grass farming from the beginning, he believed, "we should have lost a great deal less money, made a great deal more rapid progress, and saved ourselves a great deal of both heart- and back-break."[1]

When Bromfield spoke of *grass,* he was using the word in a broad sense to include several types of plants: true grasses like bluegrass, timothy, orchardgrass, and smooth bromegrass; forage legumes like alfalfa and clover; and forbs like chicory and plantain. Together, these plants provided palatable, nutritious forage that cattle could convert into milk and meat. With their dense, deep root systems, a good mix of grasses and legumes helped halt soil erosion, improved water infiltration, increased the organic matter content of the soil, and improved overall soil health. Grass could be grown on slopes too steep for safe row-crop cultivation; in fact, it was the best possible soil conservation crop that could be grown in Ohio. Properly managed grass, the "great healer," could rebuild soils faster than any other kind of vegetation, even the native deciduous forest. As an added bonus, there was a huge demand for grass-based animal products—especially milk—immediately after World War II. Grass was good for the soil, and it was profitable, too; converting Malabar from a "general farm" to a grass-based dairy farm was the best option both ecologically and economically.[2]

While Bromfield was one of the best-known advocates of grass farming in the United States in the late 1940s, he was by no means the only one. In the immediate postwar years, the USDA promoted grass farming as the best way to attain a "permanent agriculture," focusing its entire 1948 agricultural yearbook on grass. While grass had always been an important part of cattle production, Americans had not historically managed their grasslands very well. When pioneers first saw the Great Plains, they were awed by the "seemingly endless expanse of grass." But by the 1930s, overgrazing and poor range management had caused serious soil erosion and the near eradication of some native prairie grass species. In Ohio, the deterioration of pastures wasn't quite so dramatic, but there was certainly room for improvement. Ohio farmers usually selected the most worn-out, inaccessible, and rugged pieces of property they owned to be "permanent pasture," and they got out of the land about what they put into it—almost nothing. But by the 1940s, grass was finally being recognized as a crop that produced best under good management. Each major geographic region of the country had its own advocates for grass farming, and Bromfield knew them all and contributed introductions to a large percentage of the books about grass published during this period.[3]

One of Bromfield's grass-farming friends was Channing Cope, a Georgia farmer who developed a system he called "front porch farming." Cope's restoration plan for southern soils was based on four main erosion-controlling crops: sericea lespedeza, ladino clover, Kentucky 31 fescue, and kudzu. "These perennials and the livestock work for the front porch farmer, day and night, year in and year out," Cope wrote. "His only task is to *control* them." Though he waxed eloquent about many forage crops, his absolute favorite plant for soil restoration was the "miracle vine," kudzu, which was originally imported from Asia and could grow nearly a foot a day and over a hundred feet in a single year.

Cope promoted kudzu on his daily radio show and founded a Kudzu Club, which had nearly three thousand members by 1945. To Cope and his fellow "kudzu addicts," this "lovely, leaping" vine symbolized "the guiding central principle of soil conservation—*cover.*" As a legume, kudzu was great forage for cattle, and overgrazing was one of the only things that could kill it. While a few people even then were concerned about kudzu's extraordinary growth rate and propensity to smother and choke out native vegetation, it couldn't grow fast enough to suit Cope and other kudzu fanatics.[4] As local folk singer Ollie Reeves wrote in his "Song of the Kudzu Vine":

> Ah, you may have watched the black snake run
> To the shaded hole from the blistering sun . . .
> And you have seen the swallow's flight,
> And the shooting star in the deep dark night,
> But until you've watched Kudzu grow,
> You never have seen the fastest show.[5]

Kudzu couldn't handle the cold Ohio winters, so Bromfield never tried to grow it at Malabar Farm. He did, however, get excited about another plant that is now considered an invasive species—multiflora rose. The Soil Conservation Service widely promoted this tall, thorny hedge plant during the 1940s and early 1950s as a living fence or hedgerow between fields. "Compared with drab wire or wooden fences, a living fence of multiflora rose is a thing of beauty," a USDA bulletin proclaimed. Multiflora rose had attractive foliage, fragrant flowers, and colorful red berries. Its dense hedges could keep in any kind of livestock, were inexpensive to plant and maintain, provided wildlife habitat, and helped prevent soil erosion. The best varieties for planting as hedges grew to be about eight feet high and didn't spread too much horizontally if cattle grazed up to the edge of the hedge. By 1951, multiflora rose was "the keystone of cover improvement projects in Ohio and other midwestern states." Bromfield eventually planted multiflora rose on almost all of the fencerows at Malabar; it became one of the farm's distinguishing features. "At Malabar we have followed the practice of planting them along an old fence that is going bad," Bromfield explained. "We simply forget the fence and by the time it has rotted out we have a fine new hedge in its place."[6]

When Bromfield decided to specialize in grass farming, he increased his dairy herd from twenty-two milking cows in 1947 to around fifty, mostly Holsteins, in 1950. With dry cows and heifers, the dairy herd contained about a hundred animals. Although Bromfield also kept a beef herd of around fifty animals, dairy production was the main emphasis at Malabar in the early 1950s because it was highly profitable. In the im-

Louis Bromfield with multiflora rose, 1940s. (Malabar Farm archives, courtesy of Ohio Department of Natural Resources)

mediate postwar years, per-capita fluid milk consumption was at an all-time high, and so were milk prices. Between 1946 and 1951, Ohio farmers were making between thirty-two and forty-two cents of profit on every hundred pounds of milk they produced. It was a golden age for American dairy farmers—and the golden age of Malabar Farm. At the peak of this milk boom, Bromfield wrote two books promoting grass farming. The first, *Malabar Farm,* was published in 1948 as a sequel to *Pleasant Valley* and interspersed technical farming chapters with a "Malabar Journal" describing the hectic day-to-day life at Malabar. *Out of the Earth,* published in 1950, was written as a practical guide for farmers and gave a detailed description of Bromfield's farming methods and philosophy.[7]

To turn Malabar into a modern dairy farm, Bromfield had to update his buildings and milking equipment to conform with new dairy regulations that had not existed in his childhood. While Bromfield had fond memories of hand-milking cows on his grandfather's farm, the

American milk production and distribution system at the beginning of the twentieth century had serious flaws. Dirty cows, many of them infected with bovine tuberculosis, were milked in dirty barns with dirty hands into dirty milk pails. Milk was not always refrigerated on its trip to the city, where it was often watered down and adulterated by unscrupulous middlemen and dispensed to consumers out of open bulk containers. Many infants died when their mothers fed them this contaminated cow's milk instead of breastfeeding.[8]

During the Progressive Era, at the turn of the twentieth century, this dangerous urban milk supply became a target for sanitary reform laws. A New Jersey doctor named Henry L. Coit developed a system for producing clean "certified" milk in 1892. Certified milk was produced by milking healthy cows in clean barns, keeping the milk clean and cool during transport, and delivering it to consumers in sterilized glass bottles. Many doctors agreed that certified milk was safe, but it would have taken a lot of time and money to get all milk producers and distributors to comply with such stringent sanitary practices. A quicker and easier way to make the milk supply safe for urban consumers was to pasteurize the milk—heat it to kill disease-causing bacteria. By the time Louis Bromfield started producing milk in the late 1940s, pasteurization was almost universal. As one dairy scientist explained in 1948, "Without pasteurization, our dairy industry would not have grown to its present size, and our people would have been deprived of one of their most valuable foods."[9]

In the 1920s and 1930s, sanitary reformers worked to clean up the fluid milk supply chain so that pasteurization was just a precaution, not a coverup for dirty milk. To comply with new sanitary regulations, farmers ripped out old wooden stalls and replaced them with metal stanchions and concrete floors. While Louis Bromfield certainly didn't want to produce unsanitary milk, he was a "pioneer" in challenging the stanchion barn system on the grounds of "cow comfort"—animal welfare. Bromfield believed an important part of farming was "the capacity to imagine one's self as a head of lettuce, a cow, a chicken, or a cubic foot of soil and then conceive what would be best for the particular plant, animal, or cubic foot of soil." And when he imagined himself as a cow, he decided that living in a stanchion barn would be very unpleasant. "We asked ourselves how we should like being locked into a stanchion unable to move about or to lie down except in a fixed and cramped position on cold concrete for four to five months of the year," he wrote. "I have seen some expensive, plushed-up stanchion barns which might well invite investigation by the Society for the Prevention of Cruelty to Animals."[10]

Instead of keeping his cows locked in uncomfortable stanchions on cold concrete floors, Bromfield remodeled his barn to use a different system—loose housing, which was also called "pen stabling" or the "pen barn" system. In loose housing, cows could move freely between feeding and loafing areas, socialize with other cows, and rest and ruminate on warm, deep bedding that kept their udders clean. Since the cows were milked in a separate, concrete-floored milking parlor, the actual milking was done under even cleaner and more sanitary conditions than in a stanchion barn. The main disadvantages of loose housing were that it required more bedding and floor space, but Louis Bromfield believed the pros far outweighed the cons. In his loose housing system, Bromfield's cows gave lots of milk and had no mastitis. The cows at Malabar had access to the outside all year round and could eat or ruminate whenever they chose. As a bonus, the deep bedding system allowed the manure to compost in the barn, creating a superior fertilizer for the fields. The Malabar cows were bedded down on three feet of straw and a foot of sawdust, which allowed them to be "clean and comfortable at all times." "Our cows are happy cows," Bromfield observed proudly.[11]

The Malabar cows were milked with milking machines twice a day in the "spotlessly clean" milking parlor, and the milk was stored in ten-gallon metal milk cans in a cooler. The Borden Company picked up these cans once a day and took the milk to its Mansfield dairy plant for processing. In the 1940s, milk was still a very local product. Each major city had its own "milkshed," a roughly circular geographic area from which the city's milk supply was drawn, and each of the twenty-two milksheds in the state of Ohio had its own set of sanitary regulations. Malabar was

Milk was still picked up from the farm in ten-gallon cans in the 1940s. (box 69, folder 1494, Louis Bromfield Collection, Ohio State University, courtesy of Ohio Department of Natural Resources)

Cows on Mount Jeez after restoration. (box 69, folder 1496, Louis Bromfield Collection, Ohio State University, courtesy of Ohio Department of Natural Resources)

in the Mansfield milkshed, and all the milk produced there in the 1940s was processed and consumed locally. At the processing plant, whole milk was vat-pasteurized at 140–45°F for thirty minutes, cooled, and bottled. Whole milk was usually not homogenized; customers judged the quality of milk by the thickness of the cream layer in the one or two-quart glass bottles that were delivered to their doorsteps three times a week. Bottled milk wasn't delivered to anyone living at Malabar, though; Bromfield kept back enough fresh milk to supply everyone living on the farm with raw milk and homemade butter. While he wrote very little about raw milk, Bromfield was confident that the milk produced by the healthy, happy cows at Malabar was safe to drink raw.[12]

Though milk was the main product sold from Malabar, Bromfield didn't spend much time in his books writing about it. The emphasis was all on grass; livestock were just the "factory" that processed grass into saleable products. The grassland at Malabar was divided into two major categories—permanent pastures of bluegrass and white clover, and temporary meadows or hayfields with a mixture of alfalfa, brome grass, and ladino clover. On twenty-eight acres of permanent pasture near the dairy barn, Bromfield practiced rotational grazing during the summer. He divided the pasture into three eight- to ten-acre paddocks, moving the cows off a paddock when it was "moderately pastured off" and not letting them graze it again until it had "from a foot to eighteen inches of luscious new growth." Bromfield estimated that rotating pastures increased the carrying capacity of the land by a third and "greatly increased our milk production per head." He also experimented with harvesting "lush and fresh and juicy" green forage and feeding it directly to his animals during the "summer slump" to keep his pastures from being overgrazed.[13]

Even the best-managed pastures in Ohio, however, couldn't produce enough grass to graze cattle during the winter months. It was always necessary to preserve forage for winter, and, like every other Ohio farmer, Bromfield made part of his grass into hay every summer. When the hayfields grew tall, the workers at Malabar cut the forage with the farm's two mowers, let it dry in the sun for two or three days, and then raked it into windrows with two side delivery rakes. In the early 1940s, most Ohio farmers still used buck rakes to pick up loose hay and store it in the large hay lofts with which all the tall wooden barns of the time were equipped. Louis Bromfield was one of the first farmers in his region to use a pickup baler to bale his hay in small, square bales that weighed about fifty pounds each. Making square bales was expensive because of the high amount of labor involved, even though most of Bromfield's hay-baling crew were high school or college boys who were temporary summer help at Malabar. But baled hay was easier to transport and took up less space in the hay mow than loose hay, and Bromfield thought it resulted in a higher quality product.[14]

Bromfield quickly discovered that grass in Ohio grew phenomenally fast in May and June, but the three- or four-day windows of dry sunny weather necessary to properly cure hay were few and far between. While he waited for a dry spell to cut hay, the grass began to go to seed, making it less nutritious for animals. If it went too long, first-cutting hay was too poor of quality to feed to dairy cows. At one point, Bromfield even considered just using his first-cutting hay for bedding instead of straw. Eventually, he discovered that it made more sense to turn his first

A wagon of freshly baled hay at Malabar, pulled by a Ferguson tractor. (box 69, folder 1494, Louis Bromfield Collection, Ohio State University, courtesy of Ohio Department of Natural Resources).

cutting into grass-legume silage, which only needed to wilt in the field for a day or two after being cut, and bale the second and third cuttings later in the year when the weather was drier.[15]

For over half a century before Bromfield arrived at Malabar, farmers had been cutting green cornstalks and packing them in airtight silos. In the absence of oxygen, lactic acid bacteria fermented the cornstalks into a sweet-sour-smelling product called *silage,* preserving them for use later in the winter. By the early 1940s, many farmers were making silage from grass and legumes as well. After some experimentation, Bromfield concluded that grass-legume silage was cheaper to make than corn silage and increased milk production—plus, grass was much better for the soil than erosive corn. While grass-legume silage was widely considered harder to make than corn silage, Bromfield discovered that the secret to success was to make sure the silage had the proper moisture content. He tested this by making the chopped grass into a ball. If liquid came out when the silage was squeezed, the moisture content was too high; if the ball fell apart, it was too dry; "but if it makes a nice ball and you can't squeeze juice out of it," it was "just right."[16]

At first, Bromfield acquired seven small cylindrical tower silos to store his silage. Although they were an iconic part of the American rural landscape, Bromfield disliked his tower silos because they required a lot of labor to fill and were dependent on mechanical blowers, "which by breaking down could block all operations for long periods of time." When he expanded his dairy herd and needed to construct a larger silo in 1946, Bromfield opted for a totally different design—a trench silo. Trench silos, popular in the 1940s, were constructed by "making a rectangular excavation in a hillside or bank." Bromfield's trench silo, built just uphill from the main dairy barn, was fifteen feet wide, eighty feet long, and ten feet deep. He lined the sides with concrete to make it durable and "covered the bottom with about a foot of gravel." The finished silo could hold up to five hundred tons of silage, comparable in capacity to a large concrete stave tower silo. It was much easier to fill than the old tower silos—the Malabar workers simply loaded forage onto wagons with a forage chopper, backed up to the top of the hill, and dumped the forage down into the silo. A tractor-pulled scraper made it easy to pack down the silage.[17]

Hillside trench silos and their flat-land relatives, bunker silos, had many advantages over tower silos. They took far less time to fill—Bromfield estimated that it took at least fifteen or twenty minutes to empty a wagon into a tower silo, while it took only two or three minutes to dump a wagon into the trench. Trench silos were cheaper to build than tower silos, too; a farmer could build one himself, whereas most tower silos were constructed by dealers who charged a premium for their services.

Aerial view of Malabar, circa 1953, showing Bromfield's trench silo (roofed structure on right) and multiflora rose hedges in fields. (Malabar Farm archives, courtesy of Ohio Department of Natural Resources)

The only major disadvantage of a trench silo was that it exposed a larger percentage of the silage to the air. In the 1940s, agronomists suggested that farmers cover their silage with weeds, sawdust, limestone, earth, or "heavy fiber-reinforced waterproof paper." Sheets of heavy-duty "vinyl plastic" for covering horizontal silos were not available until the mid-1950s, so Bromfield solved the problem of spoilage by covering his trench with an aluminum shed roof. He piled three thousand bales of hay on top of the silage under this roof, which kept air from getting to the silage and prevented it from spoiling. In the winter, the hay and silage were removed and fed simultaneously. Bromfield was very pleased with his trench silo. "It is there not for one but for two or three lifetimes," he wrote proudly in the early 1950s. "I can report that the operation is a great success from all points of view."[18]

By 1950, Bromfield could say that the only off-farm cattle feed he had purchased in two years was some bone meal as a mineral supplement. He grew some corn, mostly on flatter land rented from the Muskingum Watershed Conservancy District, but only fed a maximum of five pounds a day, to his lactating cows. Malabar cattle got the bulk of their nutrition from grass, hay, and grass-legume silage. "At Malabar we are in one of the finest grass countries in the world and in grass and legumes lies our best hope of survival and prosperity and high profits," he wrote. Bromfield believed grass farming was an integral part of what he called the "New Agriculture," which worked with nature to increase both the quantity and quality of crops. To succeed in the New Agriculture, Bromfield explained, a farmer had to care for his soil and be "part businessman, part specialist, and part scientist." Farmers who exploited their soil would be "liquidated," and their land would go to better farmers who took good care of the soil. The New Agriculture, Bromfield told an Iowa Friends of the Land chapter in 1949, was "truly a fabulous world." "We went into grass farming because it was the cheap thing to do," he said. "And after we got into grass farming, we found our profits zooming upward." Agriculture, Bromfield explained, was no longer just about farming practices. "It is a profession today, it is a business, and it requires of the farmer to know more about more things than any other man in the world," he concluded. "I think the future of the farmer is a very great thing."[19]

American agriculture was changing rapidly in the 1940s, and farmers like Bromfield hoped that the current burst of prosperity would continue indefinitely into the future. Ironically, the war had helped American farmers economically more than any of the New Deal farm relief programs. With Europe's industrial and transportation infrastructure in ruins and global trade interrupted, American farmers were urged to pull out all the stops and produce food at peak capacity as a patriotic duty. The Office of Price Administration controlled both military and civilian food supplies during the war, and some staple foods—like meat, butter, and sugar—were rationed and sometimes unavailable. Wartime food regulation profoundly shaped American dietary habits and set the course for the industrialized food system that rose to prominence after the war. For example, military planners discouraged production of regional or ethnic foods, only supplying tin cans and other critical materials to large processing companies that produced standard "American" fare. Wartime propaganda encouraged citizens to view consumption of processed foods as patriotic and somehow analogous to American freedom.[20]

One shift in dietary preferences that Louis Bromfield tried valiantly to prevent was the displacement of butter with margarine. Oleomargarine was first developed by chemists in the 1880s as a way to make lard or tallow have the same texture as butter, and dairy farmers bitterly attacked it from the beginning. In 1902, the dairy industry successfully lobbied Congress to put a ten cent–per–pound tax on colored oleomargarine. By 1940, margarine was made from hydrogenated vegetable oils and was cheaper than butter, but Americans still preferred real butter, consuming over seventeen pounds of butter and only two and a half pounds of margarine per capita. But when the Office of Price Administration took over the food supply, it rigged the rationing system so that consumers couldn't buy butter and meat in the same week but had enough points to get both meat and margarine. By 1947, butter consumption was down to eleven pounds per capita and margarine consumption had doubled from prewar levels. Bromfield wrote some passionate articles about the danger of replacing butter made from soil-conserving grass with margarine made from soil-depleting soybeans and cottonseeds, but despite his efforts, the tax on colored margarine was repealed in 1950. By the mid-1970s, butter consumption had plummeted to four pounds per capita, while margarine consumption rose to nearly twelve pounds per person.[21]

Another major impact of wartime rationing was to drive home the idea of global food scarcity to an extent that mere propaganda never could have achieved. Not that Americans were ever in danger of starving, but the very thought that they might not be able to buy as much of certain foods as they wanted was sobering. Many people realized that postwar food shortages in Europe were temporary but wondered if American farmers could produce enough food to make up the difference. And then there was the looming question of what was first called the "Third World" in the 1950s—the former European colonies that gained their independence after the war and subsequently wanted to raise their living standards. Even before the war ended, the 1943 United Nations Conference on Food and Agriculture, held in Hot Springs, Virginia, addressed the problem of world hunger. Two years later, the United Nations' Food and Agriculture Organization was officially established at a meeting in Quebec. Sir John Boyd-Orr, a humanitarian Scottish nutritionist who became the organization's first director general, believed that improving the diets of the world's poorer nations would solve two problems—malnutrition in developing countries and surplus agricultural products in industrialized countries. He proposed creating an international "World Food Board" to equitably distribute the world's food supply where it was most needed, but his plan was never implemented.[22]

While Boyd-Orr was confident that the world could produce more than enough food to give everyone an adequate diet, others were more pessimistic. What if the war didn't cause the food shortage at all, but it was actually the other way around? What if Germany and Japan had become aggressive because their populations had outstripped their food supplies? For the first time in American history, people in the United States started to wonder if the theories of Thomas Robert Malthus, first proposed in his 1798 *Essay on the Principle of Population*, might be true after all. Malthus (1766–1834) did some math and calculated that the human population was capable of increasing in an exponential ratio. Agricultural production, he believed, could only increase linearly—and thus population would inevitably outstrip food supply, leading to war, poverty, disease, and famine. Malthus's theories were revived in the immediate postwar years, and *neo-Malthusianism*—the belief that the only way to prevent massive starvation was to implement a strict birth control program to keep the birth rate even with the death rate—became a hot topic in the United States.[23]

In 1945, the population researchers Guy Irving Burch and Elmer Pendall published an influential book, *Population Roads to Peace or War*, in which they argued that the world was already overpopulated. They warned that overpopulation led to scarcity of natural resources and food, which led to hunger, which led to political instability, which led to war—a chain of events that historian John Perkins has called "population–national security theory." Based on a misinterpretation of a study done by the Agricultural Adjustment Administration in 1935, Burch and Pendall believed that at least two and preferably two and a half acres of arable land were required to produce "a minimum adequate diet" for one person. The most optimistic estimate of the total amount of arable land in the world was 4 billion acres, so they believed that the optimum world population was somewhere around 2 billion people—half a billion less than the 2.6 billion people already on earth.[24]

By 1948, neo-Malthusianism had become a major part of the conservation movement—to the dismay of Catholic conservationists, who wanted to be good stewards of the environment but were morally opposed to "sterilization and artificial contraception." After Guy Irving Burch gave a talk at Friends of the Land's 1946 soil and health conference, Russell Lord devoted many pages of *The Land* to both sides of the population debate. The discussion reached a peak in 1948, when two major books on the subject were released: *Our Plundered Planet* by Fairfield Osborn and *Road to Survival* by William Vogt. Osborn's book was fairly typical of the stance other conservationists took at the time. While he didn't think the world could support more than 3 billion people and considered birth control a necessity, Osborn believed there might be some hope if humans started working with nature instead of arrogantly trying to dominate it. William Vogt took a much more pessimistic view, arguing that all conservation efforts would be futile until the human population was stabilized and eventually decreased by birth control.[25]

Ollie Fink, as Friends of the Land's executive secretary, incorporated a significant amount of neo-Malthusian rhetoric into the conservation speeches he gave across the country. In one of his speeches, "Democracy and Human Freedom are Products of Fertile Soil," Fink argued that a democratic government was only possible in a nation with at least two acres of arable land per person. "As acres become too few and people too many, the government steps in and regulates the distribution of food," Fink warned. "We recognize the type of government as socialist—

fascist—or communistic." "Facts rolled off Ollie's tongue like hammer blows, crushing the castles in air we of America build about safety, security, freedom from want and fear in a half-starved world of teeming millions," one listener observed. "Before Ollie was half thru, the hopeless picture he painted of the inevitable fate of the world weighed heavily upon the audience and took much of the fire from his own talking." Though Fink did mention that increased agricultural yields might help postpone ultimate disaster, the overall impression he left was that "it is just a matter of time—or maybe the atom bomb is a blessing in disguise, and the mass extermination it forebodes is not such an evil after all."[26]

At first, Louis Bromfield also took a pessimistic outlook on the world's ability to feed an expanding population, even including a chapter called "Malthus Was Right" in his 1948 book *Malabar Farm.* But by 1949, Bromfield was feeling much more optimistic about the future of feeding the world. "The world is not feeding itself and actually is on the downgrade, but it could feed itself if it chose," he wrote in a July 1949 article for the *Atlantic.* "The truth is that during the past generation a great revolution has been going on in American agriculture, which until recently has passed almost unnoticed." A combination of fertilizer, increased soil organic matter, and an emphasis on soil biology had great potential for increasing agricultural yields—just like Bromfield had done at Malabar Farm. "If what we already know were simply applied to all the agricultural land of the world and the problem of proper distribution were given consideration, the world could feed itself well," he explained. "We don't have to starve." The solution, Bromfield argued in his 1950 book *Out of the Earth,* was "to raise *more food* and *improve distribution* and *destroy waste.*"[27]

Bromfield was not alone in this optimistic outlook. In a January 1949 article for *Scientific Monthly,* Merrill K. Bennett, from the Food Research Institute at Stanford University, pointed out the fallacies in Vogt's overly pessimistic predictions. Hugh Bennett estimated that soil conservation measures were capable of increasing crop yields by 20 to 100 percent in just a few years, and he believed that the growing world population could easily be fed if soil conservation measures were applied to all the land in the United States by 1970. "If we are bold in our thinking, courageous in accepting new ideas, and willing to work *with* instead of *against* our land, we shall find in conservation farming an avenue to the greatest food production the world has ever known," Hugh Bennett proclaimed. Other soil scientists agreed that good agricultural production methods, if widely applied, could greatly increase crop yields over their prewar levels—making it possible to adequately feed the world for many years to come. The wildest optimists thought that if traditional agricultural products were ever in short supply, technological advances could grow algae and yeast in bulk to make up the deficit. Even without such techno-food options, it was already becoming evident by 1950 that good farming methods could dramatically increase agricultural production.[28]

At Malabar, Bromfield promoted grass farming as an integral part of the New Agriculture that could produce abundant food and bring prosperity to farmers all over the world. International visitors from many countries visited Malabar in the late 1940s and early 1950s, hoping to improve the agriculture of their own nations by following American soil conservation methods. Each summer from 1949 to 1953, a group of thirty international high school students sponsored by the American Field Service stopped at Malabar on their monthlong agricultural tour of the United States. Bromfield entertained the European and South American teenagers with a classic American hot dog barbecue atop Mount Jeez. Meanwhile, the number of American visitors to Malabar also increased. Garden clubs, veterans, teachers, beekeepers, and innumerable other organizations from near and far arranged field trips to Malabar Farm. One group of FFA members and farmers from Saginaw, Michigan, chartered a Greyhound bus for a two-day field trip to Malabar. After seeing the pastures, cows, and soil restoration work, they left "with the satisfied feeling that Bromfield's advance statements were true and that we had met one of America's great men."[29]

"The world will never starve to death," a newspaper reporter from Zanesville concluded after listening to one of Bromfield's stirring speeches. "The good earth can and will fulfill our needs today and tomorrow, if man will give a little to the earth."[30]

*Chapter 5*

# Conservation at a Crossroads

Late in the afternoon on Sunday, November 9, 1947, 167 conservationists climbed aboard the sixteen-car Soil Conservation Special train for a weeklong tour of Texas. They had just listened to two days of stirring speeches at the seventh annual meeting of Friends of the Land in Houston, concluding with a gigantic Texas-style barbecue, and now they were off to see examples of good and poor soil management across the state. The two-thousand-mile trip was the brainchild of Clarence M. Malone, vice chairman of the Second National Bank of Houston, who was so concerned about the negative financial impacts of soil erosion that he started an agricultural department in his bank. Burlington Lines partnered with the bank to put together a "splendid private train," where each conservationist had "an assigned compartment or stateroom." Each day followed a general pattern of three meals a day on the train, bus tours of the area in the morning, and lectures by Louis Bromfield in the afternoon. The train traveled from one location to another overnight, and the passengers would wake up in the next town to start the tours and lectures all over again.[1]

This once-in-a-lifetime tour highlighted both good and bad soil management practices all across Texas. Near Corsicana, the group saw a thirty-foot-deep gully that had washed countless tons of topsoil into the rivers, creating a huge siltation problem. Members were told that all of the farmers in the watershed wanted to practice soil conservation—except for one. After the tour left, this recalcitrant farmer went into the local Soil Conservation Service (SCS) office and said, "I'm darned tired of having them point out my place to visitors as a good example of gullies. What about getting out to my place and seeing what I need, and let's get to work on it?" In the Dallas–Fort Worth area, they visited

The Texas Soil Conservation Special train, November 1947. (box 71, folder 1504, Louis Bromfield Collection, Ohio State University, courtesy of Ohio Department of Natural Resources)

the ranch of Charles H. Harris, who was restoring native prairie grasses with good grazing management. At Amarillo, they toured the factory that made Graham-Hoeme chisel plows and saw them in action at the nearby Amarillo Conservation Experiment Station. They saw more conservation tillage near Lubbock, then arrived in Wichita Falls, where the locals were especially welcoming and gave Louis Bromfield a five-gallon hat and embroidered cowboy boots. After a final tour and barbecue in Abilene, the group boarded the train one last time and ended its trip in Fort Worth on Sunday morning, November 16.[2]

The Texas soil conservation tour was indicative of how popular and influential soil conservation had become by the late 1940s. Friends of the Land was doing great as an organization, with 8600 members in 1948, huge volumes of *The Land* topping five hundred pages each year, and a quarterly "field report" called *The Land Letter,* edited by Jonathan Forman. The *Land Letter* kept members updated on conservation news, provided a forum for both sides of controversial issues, and gave detailed coverage on the society's meetings and soil and health conferences. In August 1949, the *Land Letter* switched from an unwieldy newspaper format to a tidy, modern-looking news magazine featuring a black-and-white conservation photo on its green cover. In 1950, the name was changed to *The Land News,* and for their five dollar annual membership fee each Friends of the Land member received four issues of both magazines each year. *The Land,* still edited by Russell Lord, was promoted as a "countryman's journal." While it contained some articles about current conservation issues—such as water and air pollution and splash erosion—the majority of articles were excerpted chapters from books, book reviews, profiles on conservation leaders of the past, and fiction and nonfiction stories about country life.[3]

"We stand as of now at a very important milestone in our existence," Ed Condon told the board of directors of Friends of the Land in 1950. "We will, in my opinion, either go ahead from this point to our rightful position of influence and service to the cause, with materially increased membership and financial support and a program of positive action, or we will follow the inevitable downhill slide of so many non-profit organizations—some of them in our own field—to complete ineffectiveness, if not actual extinction." Condon and other leaders of Friends of the Land were confident that the organization would succeed. After all, soil conservation was bigger and better than ever before. Not only was the conservation train tour of Texas a huge success, but the SCS was attracting national attention with a dramatic new version of the demonstration farm—"farm facelift" or makeover days. These events, which involved months of planning, thousands of dollars, and hundreds of volunteers, compressed three or four years of soil conservation work into a single jam-packed day.[4]

One farm facelift day, at Nellie Thrasher's Farm near Frederick, Maryland, started with a bang on the morning of August 18, 1948, when a new drainage ditch was created "by the simultaneous detonation of a row of

An SCS farm facelift day, exact location and date unknown. (SCS photo, box 71, Friends of the Land Records Papers, courtesy of Ohio History Connection)

dynamite sticks." Next, thirteen lime spreaders and forty-eight veterans with bags of lime created an American flag on one of the farm's fields, while a band played "The Star Spangled Banner." Heavy machinery and five hundred volunteers limed pastures, laid out contour strips in fields, dug a new pond, built a Quonset hut dairy barn, put in two miles of fence, and painted the farmhouse. Another major SCS-sponsored farm makeover was the October 7, 1948, Oklahoma Re-Run, or New Run, near Oklahoma City. At the signal of artillery fire at 8:00 A.M., a group of reenactors on horses, wagons, and buggies charged across the land in memory of the first Oklahoma Run on April 22, 1889. They were followed by bulldozers and other heavy machinery, which built a farm pond and 1800 feet of terraces. Volunteers erected fences "so rapidly that several times they trapped automobiles filled with onlookers," seeded land to grass and alfalfa, installed a new irrigation system, painted fences and barns, and constructed a new farmhouse "with an electrified kitchen."[5]

Between 1947 and 1949, SCS coordinated more than two hundred one-day farm makeovers, including the first facelift at a farm owned by African Americans, near Hillsborough, North Carolina, in March 1949. "This is not Hugh Bennett's way of getting soil conservation spread across the United States, but it is emphatically his way of dramatizing the program," explained Wellington Brink, editor of the SCS magazine *Soil Conservation.* "Millions of people are likely to get their first soil conservation lesson on such occasions. Those who attend never forget what they see." SCS leaders freely admitted that the demonstrations had a theatrical element, and of course the average farmer couldn't afford to make so many dramatic changes in one day. But over about five years, they explained, any farmer could gradually implement the same practices and boost yields while conserving soil. "The job of conservation can be done by 1970 if people want to get it done," Hugh Bennett said after the Oklahoma Re-Run.[6]

Despite the farm makeover days' popularity, responses were mixed. Some observers saw the facelifts as a desperate last attempt by SCS leaders to influence government soil conservation policy, while the American Farm Bureau Federation and some state extension services lobbied hard to get the federal government out of direct contact with farmers. Others, blinking the dust out of their eyes after the Oklahoma extravaganza, wondered if the field days were getting too big. The emphasis on big machinery, one reporter noted, made "the whole thing look like a big-farm push." Joseph A. Cocannouer, from the biology department of Eastern A & M College in Wilburton, Oklahoma, thought the emphasis on machinery and fertilizers to the neglect of soil organic matter did "more harm than good." Back in Maryland with a "dust cold" a few days later, Russell Lord thought the field days were "mighty good," but he wondered if there was some way "to make them better."[7]

Friends of the Land, Louis Bromfield, and the SCS all adopted the same strategy to promote conservation in the early 1950s: they emphasized that conservation was good business. "There will be no city and no industry unless there are farm products to sell," Friends of the Land warned in their 1951 promotional literature. "And so it is that *business and industrial leaders have a direct and vital interest in the high efficiency of American agriculture.*" "The future is assured merely by the force of economics," Louis Bromfield believed. "Farmers can only afford to observe soil conservation practices and better agriculture or they will be liquidated. With old man economics on our side we are pretty safe." Bromfield was optimistic about the future of American agriculture. He believed that a good farmer who practiced soil conservation would "comfortably survive in times of panic and in times of good prices will become a rich man" and would "be able perpetually to maintain that rugged independence and liberty which is, of all the farmer's rewards, perhaps the greatest."[8]

Louis Bromfield saw and welcomed the rise of what Harvard economist John H. Davis first called "agribusiness" in 1955. Agribusiness, as Davis defined it, involved running a farm using the same principles as any other well-managed business. With proper attention to the business side of their operations, farmers could "adjust successfully to changing conditions in terms of size, capitalization, improved technology, and essential management skills." Davis emphasized specialization, niche marketing, superior products, and direct sales as possible ways for farmers to compete in this new agribusiness world. He believed, like Bromfield, that once backward and inefficient farmers were "liquidated," those who used good business practices would continue to be prosperous and successful. The only major difference between Davis's and Bromfield's visions of the agricultural future was Bromfield's strong belief that agriculture would only be successful if tied with conservation—something Davis didn't mention. "The New Agriculture," Bromfield said, "is based fundamentally upon the principle of working with Nature rather than fighting her as so many American farmers have done in the past."[9]

Despite his positive outlook for agriculture, Bromfield's lifestyle suffered a sudden blow when his friend and secretary George Hawkins died unexpectedly of a heart attack during a trip to New York on April 9, 1948. While Bromfield said he was relieved that Hawkins wasn't critiquing his writing anymore, the quality and success of his novels declined after Hawkins died, and the Bromfield household was never the same. Louis Lamoreaux, the architect who designed the Big House, remembered that Bromfield and Hawkins would often argue with each other, but that

"Louis always listened." Hawkins "had served as an amiable buffer for the crowded hours" for Bromfield, Mansfield reporter D. K. Woodman recalled. "George was one of those extremely rare persons who is lovable even when he is angry," Hugh Bennett wrote in a sympathy letter to Bromfield. Outwardly, Bromfield acted as if nothing had changed, but his daughter Ellen believed it was the end of an era.[10]

The number of people visiting Malabar Farm continued to increase as the 1940s drew to a close, but Bromfield began to grow restless once again. With soil restoration at Malabar complete and a grass farming program firmly in place, Bromfield told reporters that the Ohio farm "no longer challenges our agricultural resourcefulness." He wanted a new challenge, a new farm to restore as quickly and gloriously as the Ohio land—and what better place to start a second Malabar than in Wichita Falls, Texas, where the people had welcomed him so kindly with a cowboy hat and boots during the Texas soil conservation tour? In 1948, the Wichita Falls Chamber of Commerce asked Bromfield if he would be interested in repeating his soil-restoring miracle on worn-out land in Texas. Bromfield visited the area in November 1948, selected a 411-acre farm about eighteen miles southwest of town, and announced in February 1949 that he was partnering with the Chamber of Commerce to start a new Malabar Farm on the site. He planned to create a model farm that would specialize in dairy farming and vegetable crops but assured concerned Mansfield residents that his permanent residence would still be at the original Ohio Malabar Farm.[11]

While "Malabar mania" swept through Wichita Falls, construction began on simple, modern buildings at the new model farm. Bromfield's

Construction in progress at Wichita Falls Malabar Farm, 1949. (box 71, folder 1504, Louis Bromfield Collection, Ohio State University, courtesy of Ohio Department of Natural Resources)

Part of one of the many crowds at Malabar, 1940s or 1950s. (box 70, folder 1498, Louis Bromfield Collection, Ohio State University, courtesy of Ohio Department of Natural Resources)

farm manager, Bob Huge, started managing the farm in January 1950 and moved into the new farmhouse in June 1951. Huge had successfully managed the dairy operation at the Ohio Malabar, but he had difficulty adjusting to farming in the drier Texas climate. By 1953, Huge was in debt and wanted to sell the farm back to its original owner. Meanwhile, the Wichita Falls Chamber of Commerce felt like Bromfield hadn't kept his end of the deal by turning the farm into a tourist attraction like the original Malabar. On May 19, 1953, they sued Bromfield for failing to restore the farm according to their expectations. Bromfield filed a countersuit in November, explaining that the land was "badly deteriorated by over-irrigation, erosion, oil pits, salt water, mesquite growth and over-grazing." "You can't expect to turn a desert into a prosperous farm in two years," he argued. Finally, on June 7, 1954, Bromfield and the Chamber of Commerce reached a settlement. The Chamber paid Bromfield and Huge $15,000 for the property—about half of the investment they had put into it. In rather poor taste, an Ohio newspaper expressed satisfaction that "Ohio has something that Texas would like to duplicate but can't."[12]

Back at the original Malabar in Ohio, things seemed to be going great. Grass-based dairy farming continued to be a profitable enterprise up through 1951, when one economist estimated that farmers in Ohio were making an average profit of forty-two cents per hundredweight on their milk. At Malabar, Bromfield told visitors, his "fat milk check" of about $15,000 a year "keeps everything going." More than 20,000 people visited Malabar each year in the early 1950s, and journalist Inez Robb described the farm as "a cross between a bee-hive, New Year's Eve at Times Square and Old Home Day at the State Fair." On Fridays and Sundays, Bromfield was "equipped with a portable microphone through which to address the hordes of farmers and rubber-neckers who arrive by the hundreds to tour Malabar." One of the largest events Bromfield hosted at Malabar in 1950 was a "Republican farmers' rally and ox roast" on September 29. Over 4,500 people attended the rally, consuming "700 pounds of roast beef, 200 pounds of fish, 100 pounds of cheese, 5,000 buns, and about 100 pounds of wieners." The Richland County Taft for Senate Committee spent almost two thousand dollars on the event as part of its campaign to get US senator Robert A. Taft reelected.[13]

Guests eating outside at Malabar; date and event unknown. (box 70, folder 1499, Louis Bromfield Collection, Ohio State University, courtesy of Ohio Department of Natural Resources)

Not all the visitors to Malabar were adoring fans. At a Friends of the Land meeting in Louisville, Kentucky, in March 1951, Bromfield dared to call Kentucky bluegrass a "noxious weed." Shocked at his audacity, the *Louisville Times* "suggested that Kentucky keep bluegrass and Ohio keep Bromfield." Characteristically, Bromfield invited the Kentuckians to visit Malabar Farm and see for themselves, and fifty of them—including Governor Lawrence W. Wetherby—flew in two chartered planes to Malabar on July 10. With a combination of "diplomacy," mint juleps in frosted glasses, and a joke that the troublesome grass on his farm was really "Canadian blue grass," which was "about five times worse than Kentucky blue grass could ever be," Bromfield soothed the angry visitors. He served a picnic lunch to the Kentuckians, and they presented him with gifts of "bourbon whisky, a sack of blue grass seed and a Kentucky ham, which the visitors said was grown on a diet of blue grass." Good humor restored, the Kentuckians and Bromfield agreed to disagree about bluegrass, though "everybody present admitted that it is superb for lawns and in mixture with other grasses."[14]

Along with welcoming both large and small tour groups, Bromfield's personal entertaining in the Big House continued unabated. "Bromfield frankly admits his preference for spacious living and that his profitable writing must continue if he is to maintain the big house at Malabar in the style to which natives of Ohio have become accustomed," a reporter noted in 1950. "Any time Louis Bromfield sits down at the table with no more than a dozen persons he feels lonely and begins summoning reinforcements," Inez Robb reported. She recalled one September dinner with twenty-four people, including Ohio governor Frank Lausche and a couple newspaper editors. "Before such a conclave ten baked chickens, a bushel of mashed potatoes, a gunny-sack full of sweet corn, a mountain of big red and yellow tomatoes on a bale of water cress and a mammoth peach shortcake under an avalanche of whipped cream disappeared without trace." Just as the guests were leaving the table, two agricultural students from Israel arrived at the farm. Bromfield didn't want to send them away, but there wasn't a bed left in the house—so he asked the young men if they would be willing to sleep out in the barn, which they were happy to do.[15]

The next big change at Malabar occurred on December 23, 1950, when Bromfield's middle daughter, Hope, married Robert Stevens. The wedding took place at Malabar Farm, and for the Bromfield family it was a much bigger affair than the more famous wedding of Humphrey Bogart and Lauren Bacall five years earlier. Just a couple weeks later, his youngest daughter, Ellen, married Carson Geld; their wedding service was held in New York City. At first, both couples stayed close to Malabar Farm. Hope and Bob purchased the adjoining Douglas farm, and

Herbert Martin, shown here with his wife in the Big House, worked as a chauffeur and server for Louis Bromfield. (Malabar Farm archives, courtesy of Ohio Department of Natural Resources)

Ellen and Carson lived in the Ceely Rose house on Malabar property. Despite the chilling murder tale Bromfield liked to tell about the Rose house, Ellen found it a peaceful, quiet place to live. "It's wonderful to live in your own house," she said. In November 1951, she started writing a weekly "Malabar Notebook" column for the *Mansfield News-Journal* about the latest events at the farm—plowing, barn cleaning, butchering, and canning crabapple jelly.[16]

The visitors continued to come—local YWCA groups, historical societies, ballet dancers, and farmers from all over the world. Many of these visitors didn't have their own cars, and the closest they were able to get with public transportation was to take the bus to Lucas, four miles from Malabar. Bromfield usually sent his chauffeur to pick up invited guests, but other visitors had to find their own transportation to Malabar. Sometimes local residents pitched in to help—like "Red" McGuire and Rhonda Myers from Lucas, who gave a man from British Columbia a ride to the farm. In gratitude for their hospitality, the Canadian sent them a gift of moose meat. The busiest day of the week at Malabar, as always, was Sunday. Ellen explained that, while other people went to church or rested on Sunday, "We get up early and brace ourselves for the fray." She told the tale of one Sunday afternoon where an eclectic combination of guests carried on parallel conversations about the best breeds of beef cattle to raise, agricultural prices, and politics. "Soon the walls were ringing with such words as 'Angus, Hereford, Brahma, prime, 35 cents, hang the government,' all to the accompaniment of Brahms Concerto No. 2 in B Flat Minor blaring forth from the gramophone." But they still heard the dinner bell over the commotion and headed over to the dining room for the highlight of a Sunday at Malabar.[17]

No matter how many people were there—and usually they numbered more than twenty—Bromfield fed them all lavishly. From "the walk-in freezer at Malabar with its seemingly endless supply of beef, lamb, pork, and poultry" came memorable meals of "fresh spare ribs and backbone," "country-fried chicken," "home-cured ham," and "legs of lamb with home-made mint jelly." For Thanksgiving dinner in 1952, the family and guests feasted on "turkey, peas, mashed potatoes, pureed chestnuts, cranberry sauce, pumpkin and mince pie." Most of the food was prepared by Bromfield's talented cook, Reba Williams (1907–2014), whose culinary masterpieces were praised by everyone who visited Malabar. But sometimes "Mr. B." liked to do the cooking himself. On

Reba Williams driving Bromfield's Jeep, 1951. (Malabar Farm archives, courtesy of Ohio Department of Natural Resources)

one of the more memorable occasions, Bromfield tried to light the oven, "and the gas exploded in his face blowing him to the other side of the room." Somehow, he managed not to drop the "huge tureen of onion soup" he was holding. A helpful guest smeared his burned face with green salve, which made him look "somewhat like a hungry vampire," but he still finished preparing the meal—dirtying about thirty-five pots and pans in the process. "Pans don't even have a chance to get cold in this house!" Reba complained.[18]

The biggest event Bromfield ever held at Malabar was the August 9, 1952, field day sponsored by *Successful Farming* magazine. "Come see for yourself!" the magazine invited readers, after printing an article by Bromfield about how he had restored his worn-out soils "to lush productiveness." Among other things, Bromfield planned to show off a reduced-tillage experiment where he had planted corn in ladino clover, demonstrate conservation tillage, show off his trench silo, and take visitors to the top of Mount Jeez. At least six thousand people arrived for the event, and "the village of Lucas experienced the biggest traffic jam in its history as cars bearing licenses from more than one-dozen Midwest states formed an almost solid line from the village to Louis Bromfield's Malabar farm." All the local stores sold out of every kind of food they had in stock. "If we had had our wits about us, the Gelds could have made a small fortune selling pop," Ellen realized later, but no one had

Cars parked at Malabar during the *Successful Farming* Field Day in 1952. (Courtesy of Gary Zimmer)

Crowds watching tillage demonstration during the *Successful Farming* Field Day, 1952. (Courtesy of Gary Zimmer)

thought to capitalize on the occasion by providing refreshments. The Richland County Red Cross mobile disaster unit showed up for the event and ended up treating sixty injuries, including a woman who severely cut her arm on a piece of farm machinery. Overall, however, everyone deemed the field day a success—though Bromfield did say later, "It was an experience I should not like to go through again."[19]

Even as things were starting to go downhill at the Texas Malabar, Bromfield dreamed of spreading sound conservation farming methods around the world. He was excited when a group of Brazilian entrepreneurs proposed starting their own version of his famous farm, which they called "Malabar-do-Brasil." Bromfield was skeptical, however, when the project's leader, Manöel Carlos Arañha, hired his daughter Ellen and son-in-law Carson Geld to oversee restoration work on the Brazilian farm. Bromfield's respect for Carson's farming abilities, already low because he was from New York City, fell even lower when Carson smashed a car, garage, jeep, and hay wagon on one of his first visits to Malabar. Ellen and Carson would have stayed at Malabar and helped with the management of the farm, but Bromfield didn't trust them. "The thing was too precious to him, too much an expression of himself, to risk its sharing," Ellen finally realized. In the summer of 1952, she started raising money for the trip to Brazil by canning hundreds of jars of peach and plum butter and other preserves in the new canning kitchen in the Big House basement. Specialty shops in New York City sold these preserves under a Malabar Farm label. The jam and jelly

Louis Bromfield (*right*) at Malabar-do-Brasil. (Malabar Farm archives, courtesy of Ohio Department of Natural Resources)

business wasn't very successful, and many customers complained that the preserves went moldy, Ellen admitted later, but just before they left she sold the pots and pans from her jam-making operation and raised enough money to last two months in Brazil.[20]

While Ellen was in the midst of preserving season, on September 14, 1952, Mary Bromfield passed away from a heart attack. Her eldest daughter, Anne, found her body in the morning, sitting on the bed where she had been reading. When Mary died, Ellen felt that the "strange, quiet pleasantness" that Mary had brought to the house was replaced by emptiness and sadness. After her mother's passing, Ellen was "at odds" with her father more and more about how to run the farm. "Rather than bring us together the land we both loved had come between us," she sadly admitted. On February 24, 1953, Ellen, Carson, and their son Stevie set off by train to Chicago and then to New Orleans, where they sailed to São Paulo, Brazil to start a new Malabar and a new life. A month later, Bob and Hope sold their farm adjacent to Malabar and moved to Virginia. "It was obvious that the Boss, as long as he could walk, talk, think and act in his noisy, expansive way, could never share his valley," Ellen wrote later. "Then, too, perhaps the brilliant light he had cast over that valley gave his children little room in which to cast lights of their own. It was almost as if by some unspoken mutual consent we owned that it was time to go."[21]

The world was changing rapidly, and despite the ever-increasing number of visitors at Malabar Farm and the huge SCS farm facelift days, things weren't the same. As the Korean Conflict ended and the Cold War strengthened, the conservation movement struggled to keep people interested. When Jonathan Forman became president of Friends of the

Land in 1952, he was shocked to discover that the organization had only seven hundred dollars in the bank and almost fourteen thousand dollars "of bills payable." He proposed "a plan of balancing our budget and living within our income." The organization's financial problems were simple but serious—it cost more than the five dollar annual membership fee to supply each member with both *The Land* and the *Land News,* with nothing left for other expenses. "One way out might have been to raise membership dues to a $10 year minimum or more," Forman admitted. "To raise dues, however, would defeat or hinder our Society's basic purpose, to expand, to grow, to reach more and more people and enroll them as actually active members of Friends of the Land."[22]

Instead of raising dues, Forman took the "drastic action" of cutting back the cost of the organization's publications by combining them into a single, hopefully less expensive periodical. In December 1952, Forman flew to Bel Air, Maryland, to meet with Russell Lord and try to merge their disparate magazines into a coherent whole. After a two-day conference, the two editors agreed that the new magazine would include a "news digest" to keep readers informed of current conservation events. Articles would focus on a variety of topics, including ecology, soil and water conservation, watershed development, and agricultural problems. A new "Home Acres" column was designed specifically to help suburban homeowners grow their own food. Forman and Lord hoped the new magazine would be so attractive that they could get enough new members to get bulk discounts at the publisher. But despite a promising first issue of the "new" *Land,* membership in Friends of the Land continued to drop. By the end of 1953, the organization had only 5500 members.[23]

Part of the problem affecting both Malabar Farm and Friends of the Land was that, in the early 1950s, the conservation movement as a whole was struggling to define itself. Superficially, the movement seemed stronger than ever before, with the general public "on the conservation bandwagon" and familiar with the word. Conservation was so politically correct, in fact, that a lot of people were adopting the word in order to push agendas that had little to do with real conservation. But a closer look at the conservation movement showed that most of its earlier charismatic leaders had retired, leaving the actual movement "small, divided, and frequently uncertain." The amount of material published on conservation education in schools was at an all-time high, but many teachers questioned whether "narrow" topics like soil conservation were relevant to urban children. While some aspects of conservation—like the wilderness preservation movement—continued to grow, the agriculturally based soil conservation movement became smaller, less effective, and fragmented. Some conservationists argued for more government support, others for less; some for more federal control, others for "decentralization" to the states. All of this infighting severely weakened soil conservation in the United States.[24]

Meanwhile, farmers were becoming painfully aware that soil conservation was not a panacea for agriculture's economic ills. The "golden age" of farm prosperity, which had begun around 1940, abruptly ended in 1953 with the cessation of hostilities in Korea. Despite Malthusian fears of world food shortages, farmers were once again faced with their prewar problems of overproduction, surpluses, and low prices. At the same time, the costs of farm inputs increased, trapping farmers in the middle of a *cost-price squeeze.* One Ohio economist calculated that the average Ohio dairy farmer was losing fifty-six cents per hundred pounds of milk in 1952. Even though Bromfield liked to consider his farm a model of efficiency, the dairy price collapse hit Malabar especially hard because of the farm's almost complete reliance on hired labor. Exactly how profitable the Malabar dairy was during the milk boom of the late 1940s and early 1950s is hard to determine. It made a monetary profit of six thousand dollars in 1951, but a balance sheet that listed farm-produced hay and feed as expenses showed a net loss. However, there is no doubt that the dairy operation was unprofitable from 1952 on. While Bromfield blamed "leaks and waste" and lazy workers for his farm's subsequent failure to break even, the reality was that Malabar struggled with the cost-price squeeze just like everybody else.[25]

As Louis Bromfield had envisioned, only the most efficient farmers could stay in business—and even they had to be careful. Economist Willard Cochrane compared postwar industrialized agriculture to a treadmill. The first adopter of a new technology, whom he called "Mr. Early-Bird," had lower production costs and therefore earned more profit than his neighbors. But when the other farmers also adopted the new technology, prices dropped and nobody made more profit than before. A "laggard" farmer who didn't adopt the new technology couldn't sell his crop for what it cost to produce and therefore went out of business—often selling or leasing his land to his more prosperous neighbors, which Cochrane called "cannibalism." By that point, Mr. Early Bird had adopted a new technology and made a bit more profit, until his neighbors adopted it as well—creating a treadmill effect in which everyone had to keep running just to stay in business. With their immediate survival at stake, it was hard to convince farmers to adopt any conservation practices that didn't increase their short-term profits—and there just wasn't time to think about permanent agriculture or the welfare of future generations. Because the SCS, Friends of the Land, and Louis Bromfield had focused so heavily on the economic benefits of soil conservation, they had trouble keeping farmers interested as the wartime farm boom busted.[26]

Even as the rest of the nation reveled in a boom of postwar prosperity, farmers and conservationists like Louis Bromfield and Friends of the Land faced serious financial difficulties. But they weren't going to give up—not yet. No, they argued, conservation was even more important than ever before. Maybe the problem, some conservationists suggested, was that they were using the wrong word. With due respect for Gifford Pinchot, who had popularized the term during the Progressive Era, Russell Lord started to think that perhaps *conservation* hadn't been quite the right word to describe the wise use of resources. "Try however we may to breath some life into it, it still conveys the picture of mummies in glass cages, or of pickles and jellies conserved in neatly labelled jars against the inevitable day of their consumption," he admitted. Beginning in the 1950s, Russell Lord, Louis Bromfield, and other conservation leaders began to emphasize another word: *ecology.*[27]

Ecology—the study of the relationship between organisms and their environment—had always been part of New Deal conservation. Aldo Leopold and other scientists introduced concepts from animal ecology into the new field of wildlife management, while plant ecology provided the scientific basis for the SCS land use planning system. In the early 1950s, however, conservationists began using the word *ecology* to describe the relationship of humans to their natural environment. Foremost in the use of this expanded definition of ecology was Paul B. Sears (1891–1990), a longtime member of Friends of the Land, head of the conservation master's program at Yale University from 1950 to 1960, and author of *Deserts on the March,* a popular 1935 book about the Dust Bowl. At first, Sears called the relation of humans to their environment *human ecology,* but he quickly discovered that human ecology was an already established social science discipline which had little to do with biological ecology other than borrowing some of its terminology.[28]

In 1951, Sears coined a new phrase to describe the study of the relationship between humans and their physical and biological environment: *Ecology of Man. Ecology of Man* was the prototype of the modern discipline of environmental science, and the term was used by several authors until it was replaced by some version of the word *environmental* in the early 1970s. Friends of the Land quickly adopted Sears's new term, lumping agricultural ecology, their ongoing nutrition conferences, soil conservation, and other emerging environmental issues like pollution under this broad heading of *ecology of Man.* As Friends of the Land discussed how to survive in the rapidly changing world of the 1950s, Louis Bromfield suggested that they broaden their scope and impact by establishing "an organization which, for the first time in history, formulated a program and an ideal under which Man, whether on his own farm, in his county, his state, his nation or in the world could live on this planet to the greatest possible development and satisfaction." As part of this plan, Bromfield proposed establishing an ecological center somewhere in northern Ohio to collect and organize "all information in the field of man's relation to his environment" and make it available for educational and research purposes.[29]

Even though his original plan to have the Ford Foundation fund the ecological center fell through, Bromfield never abandoned the idea. After Mary died in September 1952, Bromfield planned to establish the "Mary A. Bromfield Foundation" at Malabar Farm as "a permanent center for agricultural and horticultural research and the dissemination of knowledge concerning agricultural economics and the relationship between soil and health." While Bromfield didn't have the funds to make Malabar into a full-scale ecological center right away, he pointed out that "virtually all new information regarding research in the fields of agriculture, horticulture, nutrition and health today funnels naturally through the Malabar Farm offices and files." Bromfield believed ecological agriculture was the way of the future and Malabar Farm was uniquely situated to point the way to a better world. "Someday this country and the world will be forced to wake up and undertake some ecologic pattern which employs all interlocking factors," he wrote. "When it does so, there will be an immense conservation of effort, money, energy and what you will and infinitely greater progress will be made."[30]

# Chapter 6

# Vegetables on the Middle Ground

Second perhaps only to the Big House, one of the most distinctive structures Louis Bromfield constructed at Malabar was his "roadside market to end all roadside markets," next to the imposing brick house at the Niman farm ("Bailey Place") on Pleasant Valley Road. Out of a little cave in the sandstone cliff behind this house flowed a large spring, cool and clear, at a rate of two thousand gallons per hour. From the days of the earliest settlers, this spring had been diverted into a springhouse, with huge, hand-hewn sandstone troughs for cooling milk, butter, melons, and vegetables. Bromfield remembered that in his childhood the spring water had been used to fill a horse watering trough "hollowed from an enormous oak tree," with the "cold spring water dripping over the moss and rotting wood into the beds of watercress below." The horse trough was long gone by 1954, but in its place Bromfield built his own series of troughs, supported by multicolored sandstone blocks from the local area and enclosed by an open, airy pavilion. The water from the big spring bubbled up from a pipe into the uppermost trough, ran down in little waterfalls into the lower troughs, and finally disappeared into a pipe that led to the pond and watercress beds across the street. In typical Bromfield style, the roadside stand was topped with an eagle statue in the very center of the front, with vines cascading down around the edges.[1]

On July 3, 1954, the roadside stand was completed and opened to the public. The troughs, cooled with spring water, were filled with fresh vegetables grown at Malabar Farm. Excess vegetables were stored in a more utilitarian, concrete-block structure across the street that also had troughs with flowing spring water. Soaked in this cool water, the fresh vegetables stayed firm and crisp, "not the *dead* cold of the refrigerator, but the *living* cold of the spring water, gushing out of the primeval

*Above:* Louis Bromfield (*right*) in front of his new vegetable stand, circa 1954. (Malabar Farm archives, courtesy of Ohio Department of Natural Resources)

*Right:* Louis Bromfield, his Boxers, and fresh vegetables at the stand. (Malabar Farm archives, courtesy of Ohio Department of Natural Resources)

rock." Bromfield emphasized that every vegetable sold at this stand was grown in the fields just across the street, so "any purchaser can walk a few hundred feet and see for himself how they are grown and why they are top quality vegetables." The vegetables grown at Malabar included heirloom varieties and specialty crops like "tiny red and yellow" and "Italian paste" tomatoes, Bibb lettuce, white pearl onions, okra, old-fashioned varieties of cantaloupes, potatoes, bright green Pascal celery, and celeriac. There were favorite standbys like beets, asparagus, carrots, radishes, rhubarb, strawberries, and "peppery" watercress from the pond across the street, along with novelties like butternut squash and a new variety of "purple" cauliflower that made a "large head as hard as rock when in prime condition with a faint purplish green cast in color."[2]

Bromfield's roadside market was more than just an attractive and convenient way to sell fresh vegetables. It was the first step in his plan to turn Malabar into a center for agricultural and ecological research and education, providing a contact point between visitors and the Malabar staff, house guests, and sometimes even Bromfield himself. During the summer months, visitors could pick up a wide variety of industry-sponsored agricultural brochures and other educational materials at the

*Left:* Bromfield in his vegetable stand, 1955, with vegetables, calendars, books, and jams and jellies for sale. (Malabar Farm archives, courtesy of Ohio Department of Natural Resources)

*Below:* The new Malabar Farm logo, 1953. (Malabar Farm archives, courtesy of Ohio Department of Natural Resources)

roadside market. They could also purchase books about agriculture by Bromfield and others, a specially designed Malabar Farm calendar, and jams and jellies packed by the J. M. Smucker Company and labeled with a new Malabar Farm logo featuring a line drawing of the Big House. To help fund his proposed research and education program, Bromfield sold stock in a new Malabar Farm Products Corporation. He planned to start by selling jams and jellies both at Malabar and through a direct mail campaign and eventually expand the mail-order food business to include stone-ground whole wheat flour, cheese, and Malabar-raised hams. Sales of food products were "intended to provide a continuing source of funds for the foundation's activities."[3]

Whenever possible, Bromfield preferred direct corporate sponsorship for research at Malabar. He even partnered with the Monsanto Chemical Company to test its new soil conditioning chemical, Krilium—though he later denounced soil conditioners as "dubious" substitutes for soil organic matter. One of Bromfield's biggest partnerships in the 1950s was with the Reynolds Metals Company, which was looking for new markets to keep aluminum production and profits at booming wartime levels. In the 1950s, Reynolds advertised a diverse array of aluminum products, from food packaging to irrigation piping. With the concurrence of one of the biggest building booms in American history and a shortage of wood and steel, Reynolds and other companies also marketed aluminum building materials, including "lifetime" aluminum siding that never rusted or needed to be painted. Agricultural buildings seemed a logical place to use shiny aluminum siding and roofing, so Reynolds established a "Farm Institute" in Louisville in 1950 and had developed an entire line of aluminum farm products by 1953.[4]

Given Louis Bromfield's interest in anything new in farming, it is likely that these agricultural construction materials were what attracted him to the Reynolds Metals Company. By 1953, he was good friends with one of the company's executives, William G. Reynolds, and agreed to write over 120 broadcasts for a Reynolds-sponsored radio program about the latest advances in agriculture. "My primary interest is to get readable, sound, modern information to the farmer in every possible way," Bromfield explained. After several months of providing free agricultural advice, he signed a contract on January 1, 1954, to serve as an independent consultant for the Reynolds Metals Company, which paid him two thousand dollars. In this advisory capacity, Bromfield worked with Reynolds engineers in the spring of 1954 to design an aluminum hay-drying barn.[5]

The rationale for this barn, as Bromfield envisioned it, was to free farmers from their age-old dependence on sunny weather to properly cure and dry hay. People had been experimenting with drying hay artificially for at least half a century, and by the 1940s many farmers had installed mow drying systems that used large fans to force air through ducts made of boards and up into loose hay. But Bromfield thought that this method was "awkward and inefficient." Instead, he worked with the Reynolds engineers to design a system that used a fan to suck damp air out of the barn, forcing fresh air to enter through a perforated aluminum floor and come up through the loose hay. The hay-drying barn, as Bromfield envisioned it, would be almost completely automated. Wilted grass from the fields would be blown into the barn with a silage chopper, the fan would quickly dry the hay, and during the winter specially designed "self-feeding" doors on the side of the barn would be opened to allow cattle to help themselves to their feed. Bromfield believed that the hay-drying barn would pay for itself in less than two years, just from the savings in labor and baling twine. If successful, it would revolutionize hay-making for American farmers.[6]

By the end of May 1954, the Reynolds engineers had come up with a design that fit most of Bromfield's specifications. The D. C. Curry Lumber Company of Wooster, Ohio, built the prototype hay-drying barn—a modified machine shed with aluminum siding, corrugated aluminum roofing, a floor of perforated aluminum, and a large fan in one end wall—across Pleasant Valley Road from the new vegetable stand. Bromfield first tested the barn with second-cutting alfalfa in July, excitedly reporting that it was "thoroughly dried" in just twenty-four hours. After a few more tests, he determined that it worked best to wilt the hay in the field for a few hours before blowing it into the barn. Overall, Bromfield was very pleased with his new hay dryer, but the design did have some flaws. The most serious of these was that the aluminum pole building was not airtight. Air leaked in through the corrugations in the roof and the cracks around the self-feeding doors. Also, the roof had been installed incorrectly and leaked, it was hard to fill the barn evenly with a silage blower, and the shape of the barn wasn't very practical for self-feeding (someone had to climb in and pitch hay toward the doors for the cows). Bromfield also found it necessary to construct ten-foot roofed extensions over the cattle feeding area to keep the ground from turning into a muddy "quagmire."[7]

Bromfield believed these flaws were minor and could be easily remedied. In the meantime, he helped Reynolds write up promotional materials for the barn, claiming that it produced high-quality hay and would save money on baling twine and labor. He also devoted several episodes of his radio program to discussing the benefits of the hay dryer, and he received hundreds of positive responses. These promotional efforts were so successful that people were writing to Reynolds and asking for plans even before the promotional material was ready in late 1954. But by June 1955, Reynolds had lost interest in the project, had no more brochures on the hay-drying barn, and had no plans to print more. Apparently they didn't think it would be profitable to work out the building's flaws; to make the hay-drying barn really work as Bromfield had envisioned would have required more changes than

*Right:* The completed Malabar-Reynolds hay-drying barn, 1954. (Malabar Farm archives, courtesy of Ohio Department of Natural Resources)

*Below:* Self-feeding doors in action on the hay-drying barn, 1954. (Malabar Farm archives, courtesy of Ohio Department of Natural Resources)

merely sticking a big fan and a perforated metal floor into a standard pole barn. Hay drying remained too expensive for the average farmer to consider, and the concept was largely made obsolete by the almost complete transition from loose to baled hay during the second half of the twentieth century.[8]

As the hay-drying experiment showed, Bromfield was still very interested in grass farming. But as dairy farming became less profitable in the 1950s, Bromfield began to diversify his operation once again. He started raising pastured hogs at Malabar, feeding them soil-conserving alfalfa, ladino clover, and barley instead of corn. In 1955, Bromfield planned to fence in 155 acres of swampy woodland and expand his swine program. He also considered opening a cheese factory on the farm to sell his milk as a value-added product, but that idea never got past the planning stage. By far, the most successful part of Bromfield's diversification plan was his expanded emphasis on market vegetable production. There had always been a vegetable garden at Malabar, but the Bromfield family, guests, and farmworkers had consumed most of the produce. In 1951, one of Bromfield's friends and neighbors, Bill Solomon, asked if his children could sell extra vegetables from Malabar Farm. They moved a shed to Malabar and started selling "Malabar Farm Produce" to passing motorists.[9]

By 1954, the vegetable business was so successful that Bromfield constructed the picturesque sandstone roadside market to handle the increased volume of produce and sales. He signed a contract with his two English vegetable gardeners, David Rimmer and Patrick Nutt, stipulating that he would hold 51 percent of the stock in the roadside market

The first vegetable stand at Malabar, operated by the Solomon children. (box 69, folder 1496, Louis Bromfield Collection, Ohio State University, courtesy of Ohio Department of Natural Resources).

and they would split the remaining 49 percent equally. They agreed to provide the Big House with a year's supply of vegetables in return for room, board, and laundry service. The market garden was very profitable: in 1955, Rimmer and Nutt invested $493.49 in seeds and grossed $4,733.90 in vegetable sales. "From the very beginning the business far exceeded all expectations," Bromfield was proud to report. Customers came from near and far to buy the Malabar vegetables for many reasons. For one, the vegetables were always fresh and of high quality. Also, there was the novelty of the flowing spring water in the vegetable stand and the general excitement of a visit to Malabar Farm. But one of the biggest things that drew visitors to the Malabar vegetables was that Bromfield advertised that they were grown without pesticides.[10]

"In the case of vegetables and of all foods including milk, we have no liking at Malabar for consuming in our daily meals quantities, either large or minute, of poisons universally recognized as lethal," Bromfield explained. Back when he was a teenager working on his grandfather's farm, pesticides had not been a major part of Ohio farming; insects were traditionally controlled using cultural and biological methods. He hadn't used pesticides during his sojourn in France, either, and when he returned to the United States in 1939 he was surprised to see how dependent American agriculture had become on pesticides. The change had come during and after World War I, when the use of a war metaphor for insect control became increasingly common. As pesticides rose to prominence, the USDA started testing their efficacy and removing fraudulent products from the market. By the 1920s, many American farmers—especially fruit growers in the Pacific Northwest—were routinely using arsenic- and lead-based inorganic pesticides to kill common crop pests.[11]

Pest control in the United States changed dramatically during World War II, when the Swiss chemical company Geigy released a new chlorinated hydrocarbon pesticide commonly known as dichloro-diphenyl-trichloroethane (DDT). After a couple years of testing showed that the chemical was relatively safe for humans, the American military used DDT to control mosquitoes, fleas, and other disease-carrying insects in combat areas. DDT became a war hero when it was successfully used to kill lice and halt a typhus epidemic in Naples, Italy. With an immense wave of favorable publicity, DDT was released to the public in mid-1945. It was quickly followed by other, more toxic chlorinated hydrocarbon and organophosphate pesticides.[12]

Even before DDT was authorized for civilian use, tests showed that the chemical had some unusual and worrisome properties. It accumulated in body fat, was secreted in the milk of lactating animals, and killed fish and wildlife when sprayed aerially. Chlorinated hydrocarbon pesticides accumulated in the soil and some of them, especially benzene hexachloride (BHC), made vegetables taste "musty." While cases of acute human poisoning from DDT were rare, many people wondered about its long-term effects on human health. Morton S. Biskind, a Connecticut physician, had several patients who developed a strange set of symptoms when exposed to chlorinated hydrocarbon pesticides. Other physicians, like Francis Pottenger and Granville Knight from California, made similar observations and suspected that some people were more sensitive to chlorinated hydrocarbons than others and that these sensitive individuals were the ones who showed symptoms.[13]

By 1950, many Americans wanted to know if the pesticide residues in their food were high enough to cause health problems. Concern about dangerous chemicals in food had already surfaced twice in the twentieth century—once when Harvey Wiley fought for the original Pure Food and Drug Act in 1906, and again in the 1930s when a group of muckraking writers highlighted how dangerous substances were making it into foods and drugs through loopholes in the law, leading to the renewed Food, Drug, and Cosmetic Act of 1938. The postwar era brought a surge of new chemicals in foods, and many people believed a reevaluation of the safety of the American food supply was necessary. In 1950, the US House of Representatives formed the Select Committee to Investigate the Use of Chemicals in Foods and Cosmetics, chaired by James J. Delaney. Between 1950 and 1952, the Delaney Committee held an extensive series of hearings to determine whether existing laws were adequate to protect consumers from chemicals in their food, and pesticide residue was one of the major topics discussed.[14]

During the Delaney hearings, expert witnesses on both sides testified about DDT and pesticides. Biskind and others warned about the dangers pesticides posed to human health, and food processors explained that their customers were demanding uncontaminated produce. But they were far outnumbered by the entomologists, agronomists, and public health officials who argued that pesticides were "absolutely essential" for a prosperous American agriculture. "Either we control the bugs and blights—or, like other nations in history, we shall go hungry," they argued. Many claimed it would not be possible to produce enough fruits and vegetables to enable Americans to eat a healthy diet without the use of pesticides. After a little more questioning, however, they admitted that their concerns had an economic component. Apple growers from Washington State argued that if they didn't use pesticides, they couldn't compete with growers in New York for East Coast urban markets. It wasn't necessarily that there would be *no* food without the use of pesticides; farmers just thought they were necessary to succeed in an increasingly competitive market based on an unnaturally high standard of perfection.[15]

Louis Bromfield followed the pesticide debate with great interest, and on May 11, 1951, he traveled to Washington, DC, to testify before the Delaney Committee about his experience with pesticides. Bromfield was the only witness at the hearings who was specifically introduced as a farmer, and he claimed that it was possible to grow high quality produce without pesticides. It wasn't a choice between protecting human health and staying in business; with proper soil management, farmers could do both. Bromfield explained that he had virtually eliminated pesticides at Malabar Farm. And he was not the only one; he cited several other large-scale fruit and vegetable growers who were producing good produce without pesticides. Consumer demand for pesticide-free foods was huge and growing. Bromfield had personally talked to representatives from Heinz, Campbell Soup, and the A & P chain of grocery stores, who wanted to be able to place labels on their products guaranteeing that they had "never been touched by dust or spray." The only reason that such products weren't already on the market was that food processors were having trouble sourcing produce grown without pesticides.[16]

The key to eliminating pesticides, Bromfield believed, was to grow healthy plants on healthy soil, rich in organic matter and mineral fertility. Early on in his restoration work at Malabar, Bromfield read a book called *An Agricultural Testament* by the British plant scientist Sir Albert Howard (1873–1947). Howard spent most of his career in India, where he observed both good and bad agricultural practices and concluded that the healthiest soils were those kept high in organic matter by fertilizing them with organic wastes. Building on research on soil organic matter and decomposition done by the American scientists F. H. King and Selman A. Waksman and by British scientists at the Rothamsted Experiment Station, Howard developed a practical system called the Indore Method for turning organic wastes into compost. Howard emphasized the importance of humus—the semi-stable end product of microbial decay responsible for binding soil particles together, holding water, and serving as a reservoir for mineral nutrients—to soil and plant health. He also stressed the importance of soil organisms, such as mycorrhizal fungi, in natural nutrient cycling.[17]

The central principle of Howard's organic farming method was what he called the Law of Return. A fundamental law of nature, he believed, was that the remains of all living organisms must be restored to the soil to continue the natural cycle of "birth, growth, maturity, death, and decay," which he called the Wheel of Life. Any agricultural system had to follow this law in order to be successful. Otherwise, "the farmer is transformed into a bandit." The best way to do this, Howard believed, was to compost all organic wastes and return them to the soil. He envisioned a day when all wastes from cities—including human wastes—would be composted and sent back to farms, creating a closed system of agriculture. When the Law of Return was properly followed, Howard believed, diseases in plants, animals, and maybe even humans would disappear. Pests and diseases, he argued, were nature's "Censors' Department" and only attacked weak, unhealthy plants. Based on his own experiences in India, Howard believed he had ample proof that healthy plants, fertilized with compost, were not susceptible to pests or pathogens. He concluded that animals and humans who ate healthy plants would also be healthy.[18]

"Possibly no one knows more of soil and the principles that effect growth and fertility and abundance than Sir Albert," Louis Bromfield wrote. Bromfield understood the value of composting but lacked the low-wage labor that Howard had employed in India. Instead, he carefully handled his manure to preserve its fertility and then mixed it into the soil with green manure and crop residues using conservation tillage. "Actually, we did no more than transfer the composting process from the compost heap into the topsoil of our fields," Bromfield explained. "We were growing crops in a living compost heap, with *no* additional costs either in time or labor." It was almost certainly from Howard's Wheel of Life that Bromfield derived his concept of the "cycle of birth, growth, death, decay and rebirth," which he referenced frequently in his farm books. When he reviewed Howard's *The Soil and Health* for the *New York Times* in 1947, Bromfield called it "an important book by a man who has few peers in agriculture and possesses a profound knowledge of the soil which is the foundation of our existence." In return, Howard read *Pleasant Valley* and called Malabar "an example of well farmed land for all to see and copy."[19]

As much as Louis Bromfield respected Albert Howard, he disagreed with him about one thing: chemical fertilizers. These were originally invented, long before Howard coined the phrase, in an attempt to satisfy the Law of Return. Back in the 1840s, the German chemists Karl Sprengel and Justus von Liebig discovered that plants removed mineral elements from the soil, emphasizing that it was up to the farmer to satisfy "a natural law" by returning these elements in the form of organic wastes. In 1843, John Bennet Lawes and Joseph Henry Gilbert established the Rothamsted Experiment Station in England to see whether it might be possible to add minerals to the soil directly. After twenty years of experimentation, they concluded that wheat on plots given only mineral fertilizers seemed to be doing just as well as wheat fertilized with animal manure. Over the next fifty years, an entire chemical fertilizer industry developed to supply plants with important minerals like phosphorus and potassium; synthetic nitrogen fertilizer did not become widely available to American farmers until after World War II.[20]

For half a century, many agronomists believed that chemical fertilizers could completely replace organic wastes with no detrimental impact on plant health. But by the 1920s, the researchers at Rothamsted noticed that that their chemically fertilized plots had greater variations in wheat yield and had experienced more soil deterioration than plots fertilized with manure. Researchers like Selman A. Waksman at Rutgers University began to discover more about the intricacies of soil life and organic matter decomposition, and by the 1940s even chemical fertilizer industry publications called organic matter the "life blood of productive soils." In 1949, the American Plant Food Council and the National Grange sponsored a youth essay-writing contest on the topic of soil conservation. The grand prize—a Buick Super 4-Door Sedan—was awarded to a nineteen-year-old Nebraska farm boy named Wilfred M. Schutz, who stressed the importance of both commercial fertilizer and manure in his prizewinning essay.[21]

While Howard was developing his organic farming system in the 1930s and 1940s, mainstream agricultural science was slowly but surely moving in the direction of a greater emphasis on soil health and the importance of organic matter. But by the time Howard wrote his organic farming books, he had become critical of all official agricultural research, arguing that randomized, replicated experiments on a "pocket handkerchief" of land at an experiment station were not applicable to real-life farm conditions. He denounced the average specialized agricultural researcher as a "laboratory hermit" who learned "more and more about less and less." Howard had always regarded chemical fertilizers as inferior to organic wastes because they didn't restore organic matter to the soil, but by the 1940s he was calling them "one of the greatest follies of the industrial epoch." Chemical fertilizers would eventually cause diseases in plants and animals by somehow altering the process of protein synthesis. "Artificial manures lead inevitably to artificial nutrition, artificial food, artificial animals, and finally to artificial men and women," Howard claimed. And he wouldn't compromise. When agronomists suggested using both manure and chemical fertilizers, Howard expressed indignation that they were "seeking sanctuary in the humus bombproof shelter" and emphasized the need to "evict these unwelcome intruders."[22]

The leading American proponent of organic farming, J. I. Rodale (1898–1971), concurred with Howard about the dangers of chemical fertilizers. Rodale claimed that chemical fertilizers killed earthworms, bled the "soil to death," and killed "off in cold blood quadrillions of bacteria and fungi." Not surprisingly, advocates of chemical fertilizers believed they had to defend themselves against such accusations. While admitting that soil organic matter was important, they claimed there just weren't enough organic wastes in the world to restore all of the required mineral elements to soil and that chemical fertilizers were not only safe if used correctly but actually beneficial to earthworms and other soil life. Some scientists, like Ray I. Throckmorton from the Kansas College of Agriculture, stooped to the same kind of exaggerated language that Howard and Rodale had used, calling organic farming "bunk" and its proponents "cultists relying on half-truths, pseudo science and emotion."[23]

Louis Bromfield tried to stay neutral as he watched the "organic-chemical fertilizer feud" unfold. While he agreed with proponents of organic agriculture about the dangers of pesticides, he disagreed when they lumped chemical fertilizers into the same toxic category. At the beginning of his restoration work at Malabar, Bromfield used superphosphate fertilizer, along with lime, to establish legumes and grasses and quickly increase soil organic matter. Since he saw firsthand the same improvements in soil health that Howard had observed with his organic farming methods, Bromfield just couldn't agree that chemical fertilizers were universally detrimental. He believed that it all came down to a question of balance. On soils "virtually devoid" of organic matter, chemical fertilizers would cause the tiny amount of remaining humus to oxidize too rapidly, with a detrimental effect on soil life. But when chemical fertilizers were used judiciously to grow soil-restoring crops, they could actually increase soil organic matter and improve soil health much faster than strictly organic methods. "The answer lies somewhere between the two extreme schools of chemical and organic absolutism," he concluded.[24]

Russell Lord took a similar middle-of-the-road stance on the organic debate and tried to fairly present both sides of the controversy in *The Land,* explaining that the magazine "takes a middle position in such matters, but certainly maintains an open mind toward organiculture." When J. I. Rodale published his first major book on organic farming, *Pay Dirt,* in 1945, Lord devoted seventeen pages of his magazine to both positive and negative reviews of the book. In the 1950s, he published profiles of successful organic farms, like Alden Stahr's Stahrland Farm in New Jersey. Stahr built up his farm's worn-out soils with a combination of manure, organic wastes, and rock powders like greensand, phosphate rock, granite dust, and limestone. Stahr found that the health of his plants and animals improved dramatically under his system of "Normal Agriculture," making pesticides and medications unnecessary. In 1951 Stahr visited several other organic farms in New York, Pennsylvania, Ohio, and Michigan where people were growing grains, vegetables, and dairy cattle without any pesticides or soluble chemical fertilizers. The demand for food grown without pesticides was very high; the farmers Stahr visited sold everything they could produce, some shipping organic products across the country direct to consumers. Stahr believed natural

Louis Bromfield grew high-quality vegetables at Malabar, like these tomatoes, without pesticides. (A/V box, from box 11, Friends of the Land Records Papers, courtesy of Ohio History Connection)

farming would eventually replace chemical agriculture due to consumer demand.[25]

Sometimes in the same issue as these glowing reports of successful organic farming operations, Lord also printed anti-organic letters from writers like David Greenberg, who thought the "Organitics" were "cockeyed wrong." "I can read the works of the prophets of this organic cult and knock them out and draw and quarter the lot of them with quotes right out of their own pages," Greenberg claimed, criticizing Stahr's refusal to use soluble chemical fertilizers but admitting that he didn't even know what greensand was. In response to Greenberg's attacks, Lord published letters from organic advocates like Leonard Wickenden and Robert Rodale. Russell Lord thought the emotional arguments on both sides made for "good, warm human reading" but agreed that the debate produced "more heat than light." He tried to let both sides speak for themselves so that readers could make their own decisions.[26]

Even though he marketed his Malabar produce as pesticide-free, Bromfield continued to use chemical fertilizers in his vegetable fields. In 1953, he collaborated with the Battelle Memorial Institute of Columbus, Cliff Snyder of the Sunnyhill Coal Company, and Ollie Fink and Jonathan Forman from Friends of the Land to develop a complete soluble fertilizer for foliar application in irrigation water. The final product, which contained trace element sulfates in addition to nitrogen, phosphorus, and potassium, was marketed under the trade name "Fertileze" by a company called Nutrition Concentrates, Inc. Bromfield was very pleased with the performance of Fertileze; he traveled to garden stores to introduce the new fertilizer and endorsed it in newspaper advertisements. Bromfield found that with a combination of fertilizer, irrigation, and a system of intensive mulching, the Malabar vegetables were largely free from pests and diseases and were of such high quality that people came from miles away to buy them at the vegetable stand. He saw no evidence to support the claim that properly used chemical fertilizers killed soil life—or the counter claim that it was impossible to grow good produce without pesticides.[27]

In 1955, Louis Bromfield published *From My Experience,* his fourth and final book about his accomplishments at Malabar Farm. He wrote about his hay-drying barn, his vegetable stand, the pastured hog operation, Malabar-do-Brasil, and all the new and exciting things happening at Malabar. In his more philosophical passages, Bromfield expressed his frustration with the ongoing controversy between "the organic extremists and even the cranks" and the narrow-minded, overspecialized views of many agricultural scientists and academic institutions. Even as the organic controversy became more heated, Bromfield tried hard to maintain an open-minded, middle-of-the-road view. "At Malabar Farm we have never been *doctrinaire* nor have we been cranks," he explained.

> We are primarily interested in what works and what is permanently good for the farmer, his soils, his crops, and his animals and what gives him the highest production and the best nutritional quality. . . . Nevertheless we have our theories and these have led in a single and apparently inevitable direction—that nature has provided the means of producing healthy and resistant plants, animals, and people and that if these means and patterns can be discovered and put into use, the need for "artificial" and curative, as opposed to preventative methods, is greatly reduced.[28]

Bromfield still hoped the controversy was temporary. He hoped the two sides would reconcile their differences and work together to create a permanently prosperous agricultural system that worked with natural processes. As the 1950s wore on, however, the divide between Bromfield's vision and the reality of American agriculture became greater each year. The battle lines being drawn in 1950 hardened into place, with the adamant advocates of chemical fertilizers and pesticides on one side and the equally adamant "organiculturists," led by J. I. Rodale and his *Organic Gardening* magazine, on the other. Bromfield's balanced view was fast becoming a disputed no man's land, harder to maintain as time went on—and the dream of the New Agriculture was turning into a nightmare for many farmers.

*Chapter 7*

# Saving Malabar

On November 17, 1954, a hundred local farmers gathered at Malabar Farm, near Bromfield's new hay-drying barn and vegetable stand. That in itself wasn't unusual; after all, groups of farmers frequently visited Malabar. But these farmers were attending an auction at which Bromfield sold off fifty of his Holstein dairy cows and heifers. It was "no longer possible to profitably operate a dairy farm," Louis Bromfield told a newspaper reporter when he announced the sale. "Almost any other farm activity is more profitable and much less trouble." Like every other dairy farm in the country, Malabar was caught in the cost-price squeeze and couldn't sell milk for what it cost to produce. The only dairy farms that managed to stay in business were family-run operations, where the entire family worked long hours with no hired help. "It requires careful management and a good herd for a dairy farmer to stay in business," a neighboring dairy farmer explained. With so many other farms in the same boat, it was a bad time to sell cattle. Despite the auctioneer's best attempts, the cattle sold for "terribly low prices"—only $75 to $140 per head for animals that were worth $200 to $250.[1]

Less than a month after Bromfield downsized his dairy herd, friends and fans alike were shocked to learn that the famous author had been hospitalized in Mansfield with an "acute infection." He had recovered enough by February 1955 to visit Ellen and Carson at Malabar-do-Brasil, where they were making excellent progress restoring worn-out land to productivity and adapting to Brazilian culture. Ellen noted that her father seemed in remarkably good health for "one who is supposed to be recovering from a near fatal attack of blood poisoning" and that she was having trouble keeping up with his "break-neck speed." The visit turned out to be a very good thing for all concerned. Finally, after many

Fazenda at Malabar-do-Brasil where Ellen and Carson Geld lived. (Malabar Farm archives, courtesy of Ohio Department of Natural Resources)

years of tension, Ellen reconciled with her father. He was genuinely impressed with their restoration work and stopped criticizing everything Ellen and Carson did.[2]

It was while staying in a "white room" in the Geld's fazenda that Louis Bromfield read *Out of My Life and Thought* by the philosopher and missionary Albert Schweitzer. Bromfield was deeply affected by Schweitzer's phrase "Reverence for Life," which Schweitzer defined as "the affirmation of life and ethics inseparably combined." "The man who has become a thinking being feels a compulsion to give to every will to live the same reverence for life that he gives to his own," Schweitzer wrote. After reading Schweitzer's book, Bromfield believed that "in this very great phrase, 'Reverence for Life,' I too found what I had sought for so long." Bromfield's definition of "Reverence for Life," which was slightly different than Schweitzer's, emphasized the concept of working with nature that he had been promoting at Malabar for the past fifteen years. "Every good farmer practices, even though he may not understand clearly, the principle of Reverence for Life, and in this he is among the most fortunate of men," Bromfield concluded.[3]

Five or ten years earlier, this idea of reverence for life or working with nature was an important part of the conservation movement. But as the New Agriculture turned into agribusiness and the total number of farmers in the United States continued to decline, the soil conservation movement became smaller and less effective. Friends of the Land, so vibrant in the 1940s, continued to lose members as the 1950s wore on. The "new" *Land* magazine and plan for expanded growth in 1953 turned out to be a dismal failure. The larger page size of the combined magazine was much more expensive than Friends of the Land could afford to publish, and Russell Lord and Jonathan Forman continued to debate over the periodical's content. Lord still dreamed of an agrarian country life magazine with tens of thousands of subscribers; the more practical Forman wanted an informative news magazine that would further the cause of conservation. Things came to a head at the board of directors meeting in Topeka, Kansas on March 23, 1954. Forman told the board that continuing to publish *The Land* in its present form would put the organization in debt. "As I see it, it is not a question of publishing *The Land* or doing something else," he explained. "It is necessary that we should discontinue *The Land*." As part of their drastic cost-cutting measures, Friends of the Land also closed their Columbus office and moved the national office to Hidden Acres, Ollie Fink's farm near Zanesville, Ohio.[4]

After the board meeting, Friends of the Land officially gave Russell and Kate Lord copyright to *The Land* and agreed to purchase copies of the magazine at fifty cents apiece for their members in 1954. The magazine's affairs were turned over to a new nonprofit organization called "The Land Trustees," with Louis Bromfield, Paul Sears, Jonathan Forman, and several others on the board. Free from the demands of the Friends of the Land Board of Directors that the magazine be more practical, Russell Lord turned volume 13 of *The Land* into the country life magazine he had always dreamed of producing. It was poised to appeal to a greatly expanded audience, he told readers. But it soon became painfully obvious that, no matter how much Lord wanted it to be, *The Land* was not relevant for the 1950s. The expanded subscriber base he envisioned never materialized, only two issues were published in 1954, and the magazine's debts at Waverly Press continued to mount. On January 25, 1955, the Land Trustees met and discovered that they owed $4,647.96 to the printer—exactly the situation that Forman had urged Friends of the Land to avoid. Reluctantly but firmly, the trustees took the only step possible—"the immediate discontinuance of the magazine." "It is all very sad indeed and I only wish that I had the time and money to save the situation," mourned Louis Bromfield, who was too ill to attend the meeting.[5]

To replace *The Land,* Jonathan Forman and Ollie Fink coedited a much shorter, less expensive publication, which they called *Land and Water.* This new quarterly magazine, with a redesigned green cover and about thirty pages per issue, had a similar format to the old *Land News* but with longer articles and fewer news items. Responses to the new magazine were mixed. One reader was "shocked and disappointed" that *The Land* had been discontinued; others applauded the new magazine as "much more useful and practical" and "more helpful" than the earlier one. Despite its shorter size, *Land and Water* contained just as much—or even more—practical conservation information as the last few volumes of *The Land.* While Russell Lord thought they were "watering down" the title because he owned the copyright to *The Land,* the new name reflected an overall shift toward a greater emphasis on water conservation in the conservation movement in general. The first issue, written largely by Ollie Fink, was devoted entirely to a comprehensive overview of various aspects of water conservation. The second was a primer on soil-health relationships, and the third focused on "Home Acres"—part-time farming on suburban lots or in small communities. Friends of the Land hosted conferences on these three topics each year in the late 1950s, in addition to making annual appraisals of the conservation movement and tours of the Muskingum Watershed Conservancy District.[6]

Finally realizing that they would never be a large organization or have the magazine circulation Russell Lord had envisioned, Friends of the Land put out new promotional material portraying themselves as small but influential. Even in a democratic society, they explained, public opinion was shaped by a surprisingly small group of influential people. "One thousand lecturers and writers produce the speeches, the books, the pamphlets and the shows on stage, screen and television, which a quarter of a million teachers, columnists, preachers and commentators pass on as ideas and concepts to 165,000,000 Americans. This process is not conspicuous. You have to hunt before you can discover it. Of such is Friends of the Land." They argued that the organization's effectiveness should not be judged by the number of members or subscribers to its publications, but by the number of influential people persuaded. Just as Ollie Fink had emphasized with his conservation education programs fifteen years earlier, the goal of Friends of the Land was to teach the teachers. Unfortunately, the organization no longer had the access to influential leaders that it had enjoyed in the 1940s, and it is questionable whether it still belonged in that elite opinion-shaping group by 1955.[7]

Louis Bromfield signed his name to the first issue of *Land and Water* but was able to devote little time or energy to Friends of the Land afterward. As his health declined, he was unable to be involved in as many conservation organizations as he had been just a few years earlier. He resigned from the Ohio Wildlife Council in May 1955, "pleading the press of duties on his Mansfield farm 'Malabar' and the time consumed by his writing." Meanwhile, he planned more ways to generate income at Malabar. In June, he announced that he was going to remodel the historic brick house by the vegetable stand, built in 1820 by the pioneer David Schrack, into a restaurant and event center called "Malabar Inn." He planned to mostly use the inn for private parties. "It will be exactly like giving a party in your own home, but with the very best food and wines in the world," he told newspaper reporters. Bromfield planned to hire a French chef and considered eventually building "a series of small attractive cottages where people who dine at the inn can spend the night."[8]

The very fact that Bromfield was planning to charge people to eat at Malabar demonstrated how much he was beginning to struggle financially. He had always prided himself on feeding dozens of guests a day with the finest food and alcoholic beverages and never charging them a penny. For example, Bromfield had celebrated his own birthday in December 1953 with what he called a "lobster wallow." He had "forty-eight lobsters and three bushels of clams" shipped directly from Maine in "big cans filled with seawater and seaweed." The seafood was served with "good yellow butter" in the big sandstone-walled basement room with a fireplace so that the concrete floor could be hosed down when everyone was done. While this party might have been a bit more extravagant than normal, the reality was that he ran the Big House at Malabar Farm like a fine hotel—without charging guests for room or board. Bromfield's personal tastes ran up a considerable bill, too; he was a heavy drinker and smoker and only bought the finest liquor and cigarettes. The problem in 1955 was not that Bromfield was running out of money—he still had a net income of about $47,000 a year—but that he was, in the words of historian Ivan Scott, "probably the most improvident rich man one may read about." His later novels hadn't been nearly as popular as his early works, the farming operation at Malabar had gone from making a small profit to being a net loss, and the cost of everything had increased. Other farmers tightened their belts and put off major purchases to get by. But Bromfield abhorred the idea of thrift or frugality. He wanted to live well and be a generous host—even when "each year the expenses grew greater and greater and the means to pay them less."[9]

Bromfield's increased awareness of finances was evident in one of his last writing projects, a couple chapters for a book about his friend Charles Pettit's Flat Top Ranch. Pettit, a millionaire who made his fortune in the East Texas oil fields, decided to retire due to ill health and purchased the 17,000-acre Flat Top Ranch near Walnut Springs, Texas, in 1937. Historically, the area had been carpeted with a lush crop of waist-high grass, but it had deteriorated from overgrazing by the time Pettit took over. Pettit restored his worn-out land to productivity by stocking fewer cattle, using chemical herbicides like 2,4-D and 2,4,5-T to eradicate mesquite and other woody shrubs, and constructing forty-three artificial lakes to irrigate pastures and provide drinking water for his cattle. Bromfield first met Pettit when he visited his ranch on a tour in 1949, and by 1951 the two men were close friends. Early in 1954, Pettit corresponded with Bromfield about writing a couple chapters for a book about his restoration work (published in 1957 under the title *Flat Top Ranch*). When he was finally feeling well enough to start on the project in April 1955, Bromfield sent Pettit a letter explaining the "embarrassing" fact that he would have to accept Pettit's "kind offer of payment for the work." He would have liked to write the chapters just out of friendship, but "the account books" told him that was not possible.[10]

By the end of 1955, Louis Bromfield was a very sick man. Though he told no one except his immediate family, he was suffering from "an obscure kind of cancer of the bone marrow"—multiple myeloma. After being hospitalized in New York City, he spent several months at the New Jersey estate of the tobacco heiress Doris Duke while undergoing blood transfusions and other treatments for his condition. This resulted in a considerable amount of gossip about the possibility of a love affair between the two celebrities. "There's nothing to it," Bromfield told the nosy newspaper reporters. "We are only good and very old friends." The reporters might not have cared so much about Bromfield's romantic life if they had realized how sick he really was. By February, he had

(*From left*) Louis Bromfield, Ann Pettit, and Charles Pettit at Flat Top Ranch. (A/V box, from box 13, Friends of the Land Records Papers, courtesy of Ohio History Connection)

pneumonia on top of the bone cancer. In early March he was admitted into University Hospital in Columbus with a serious case of homologous serum hepatitis. His daughters Hope and Anne stayed in the governor's mansion with Bromfield's friend Governor Lausche and his wife so that they could be close to their father. Still, the newspapers did not realize that the famous author was dying; they reported on March 11 that he was in "satisfactory" condition. Just a week later, on March 18, 1956, Louis Bromfield passed away in the Columbus hospital.[11]

Friends and fans alike were shocked when they heard the news. Bromfield had always seemed so vigorous and full of life; the thought that he was dead was "like a bad dream." Five or six hundred people attended the memorial service at the First Congregational Church of Mansfield on March 22, and newspapers and magazines across the country published tributes to Bromfield. "I am so shocked at Mr. Bromfield's death that I cannot find any words to express my feelings," Charles Pettit wrote. "I feel like . . . the world has lost a great citizen and that I lost a great friend." "His passing leaves a lump in the throats of all of us who love the land," said Ollie Fink. "Until the very last, it seemed impossible that death could touch this man who had always been so full of life and loved it so much," Inez Robb wrote when she heard about Bromfield's death. "And now Louie has gone back to Malabar and the earth he loved so much, and those of us who loved him are infinitely poorer."[12]

As the reality of Bromfield's death set in, many of his friends began to ask the inevitable question: "What will be done with Malabar Farm now?" In his will, Bromfield left his world-famous farm to his daughters but gave the trustees of his estate "the right to make other use or disposition of the farm property." A thought occurred to Ollie Fink and Jonathan Forman—why not make Malabar into the ecological center that Bromfield had dreamed of back in 1952? Within a month of Bromfield's death, Friends of the Land started corresponding with Clyde Williams at the Battelle Memorial Institute in Columbus about the possibility of using Malabar as an agricultural research center. Established in 1929 as a memorial to iron magnate Gordon Battelle, by 1956 the laboratories of the Battelle Memorial Institute covered twelve acres of land, housed thirty-eight research divisions ranging from metallurgy to pollution control, and employed over two hundred scientists and technicians. In the Battelle philosophy of research, there were no strict delineations between different research specialties or units. Researchers from every area collaborated on projects for private industries, which provided funding and in return received the patent rights for ensuing discoveries.[13]

By the end of 1956, Friends of the Land had drawn up a detailed plan for an ecological center at Malabar Farm. It was a broad, ambitious project that would have been impressive for an agricultural university, let alone an aging conservation organization that was already struggling financially. The first objective, of course, was to purchase Malabar Farm. Apparently unaware of Bromfield's failure to make a profit before his death, Friends of the Land planned to operate the farm "as an income producing venture" to finance "practical field research and demonstrations at Malabar." Since the plan was to operate Malabar profitably, they anticipated that "visiting farmers, interested scientists and conservationists from all over the world would come to observe the results of new concepts of agriculture." At some point, they also hoped to demonstrate soil conservation methods on "a selected group of widely located, presently poorly managed, low-income producing farms."[14]

Concurrently with operating Malabar as a demonstration farm, Friends of the Land also planned to turn it into an agricultural research center, tentatively called the "Malabar Research Institute." Based on suggestions given by the Battelle Institute, they proposed an extensive research agenda that would rival programs at most land grant universities. Possible research areas included plant nutrient uptake; improved methods of soil testing; enhancing plant growth by increasing atmospheric carbon dioxide; a better understanding of the factors affecting plant maturity, yields, frost tolerance, and nutritional value; economic studies; land use planning; and watershed management. In addition to this type of basic research, the farm would continue to serve as a place for manufacturers to test and demonstrate new and improved farm machinery. Friends of the Land even hoped the program would be able to provide fellowships for international students, especially from "underdeveloped countries," to participate in "research studies and field work" at Malabar.[15]

The most distinctive part about the plan for Malabar—and, in Paul Sears's mind, the most important—was the ecological center. This center, initially located in the Big House at Malabar, would be a national clearing house for "the collection, correlation, interpretation, and information" related to the study of "Man's relation to his environment." A central part of this institution would be a "Library of Ecology," staffed by a knowledgeable research librarian who could quickly assemble "all of the information which the investigator needs for the exploration of his problems." The center would serve as headquarters both for Friends of the Land and the Ecological Society of America, which lacked a central office and expressed "definite interest" in establishing one at Malabar. Another function of the ecological center would be to host conferences and symposia on ecological topics and perhaps even provide "a summer laboratory for school teachers in Ecology," similar to the Conservation Laboratory Ollie Fink had established fifteen years earlier.[16]

Put together, it was an incredibly ambitious plan that touched on almost every area of contemporary agricultural and ecological research.

Ellen Bromfield Geld gave her approval and hoped the plan would succeed. Hope Stevens agreed that it sounded great, but she was more skeptical about the ability of Friends of the Land to actually get funding for the project. Battelle was committed enough to the project to pay Paul Sears's travel expenses to meet with representatives of major foundations across the country, but the results were disappointing. A representative from the Ford Foundation met with Sears and explained that the proposed ecological center was outside of the foundation's scope. Resources for the Future told him that the project "would fall completely out of their established program." The Rockefeller Foundation agreed about "the need for a center for unconventional research in agriculture and land use" but suggested looking elsewhere for funding. Sears tried his best but reported on March 14, 1957, that he had "'no results' though I did canvass the possibilities as thoroughly as I could by telephone and travel." Clyde Williams from Battelle thanked him for at least trying. "You put forth a fine effort on a very difficult assignment and we certainly appreciate your jumping in and taking time from your busy life." "I am sorry not to have brought home the bacon," Sears replied.[17]

Despite Sears's best efforts and personal prestige as an ecologist, no one was interested in funding the type of ecological center that Friends of the Land wanted to establish at Malabar. Perhaps their plan was too ambitious; the proposed agricultural research program, for example, would largely have duplicated research already being done at land grant universities and would have required a steady, ongoing source of substantial funds to implement and maintain. But there may have been other reasons no one wanted to be associated with the project. Ollie Fink confided to Ed Condon that a friend had "allegedly reported" that charitable foundations "have been put under terrific pressure by the so called 'regular Agricultural Agencies[']' (Agricultural Colleges, Extension Services, etc.) to stay out. The line being that Bromfield and FOL are crackpots and should be discouraged." It was true that Bromfield had criticized the overspecialization of agricultural research, though he repeatedly emphasized that he was not a "crank" or "organiculturist." Friends of the Land, however, had generally been supportive of the land grant universities and tried to remain neutral on issues like organic farming, at least while Russell Lord was editing *The Land.* If anyone called them crackpots in the 1950s, it was likely because of the increased emphasis they put on the soil-nutrition connection after Jonathan Forman and Ollie Fink took over editorship of *Land and Water.*[18]

Back when Forman and Fink first met in 1940, research on the connection between soil fertility and human health was inconclusive, leaving room for a wide range of speculations. By the mid-1950s, however, much more research had been done on the soil-health connection. In

1948, Firman Bear from Rutgers University analyzed the mineral content of snap beans and tomatoes from several different states and found that the concentration of most minerals tended to increase from east to west. William Albrecht and Jonathan Forman hailed the Rutgers findings as "incontrovertible" evidence for their hypothesis that "poor land makes poor people." But a slight variation in mineral content did not mean that vegetables grown in the East were devoid of nutritional value, nor that people eating them would automatically be unhealthy. Meanwhile, researchers at the USDA's Plant, Soil, and Nutrition Laboratory at Cornell University ran a huge number of experiments and discovered, among other things, that fertilization increased yields but did not significantly raise the mineral content of plants; there was no significant difference in growth between animals fed fertilized and unfertilized forages; and differences in the vitamin C content of tomatoes from various regions was caused by sunlight, not soil type.[19]

One of the most carefully planned, long-term "biological assays of soil fertility," as Albrecht would have called it, was conducted from 1945 to 1955 by scientists from Michigan State University. In this multidisciplinary study, researchers started with a "badly depleted farm" that they determined had not been fertilized or manured for at least a hundred years. They fertilized half of the land and left the rest unfertilized, grew the same forage and grain crops on both treatments, and fed them to two identical herds of dairy cattle for three or four generations. To the surprise of many of the researchers, there was no significant difference in the mineral content of the crops (except timothy), the health of the cows, milk production, or the health of rats raised on the milk produced by the cows. But Forman continued to cite the experiments Albrecht had conducted fifteen years earlier, criticizing the methodology of the Michigan State study because it didn't have the results he expected. In 1948, Paul Sears had described Forman, Albrecht, and several others as members of "a very lively Advance Guard which is responsible for growing popular interest in the relation of Soil to Health." By the mid-1950s, however, this Advance Guard seemed to be falling behind, stuck in the theories of 1940 and unwilling to accept that some of them might have been flawed.[20]

Whether Forman and Fink's reluctance to stay up to date in the field of soil and health research made major foundations wary of funding the Malabar project is unknown. They didn't put much emphasis on the soil-health connection in their plans for an ecological center, and most of the other issues that Friends of the Land discussed—water conservation, crop surpluses, how the cost-price squeeze was affecting American farmers, the destruction of farmland by suburban sprawl, and the need for wilderness preservation—kept up with the times. Perhaps the main problem was that the whole idea of an ecological center was too novel, too interdisciplinary and unconventional in an age of increased specialization in every area of science. For whatever reason, no major foundation was willing to help fund the project.

The executors of Bromfield's estate were open to the idea of having Friends of the Land turn the farm into an ecological center, as were many others in the local area. Two businessmen from the nearby town of Galion, Ralph and Herb Cobey, managed the farming operation at Malabar while they waited for Friends of the Land to raise the money to purchase the farm. But as months went by and Friends of the Land failed to find funding, the executors grew tired of waiting for something that might never materialize. On January 7, 1957, they listed the farm for sale with a real estate agency for $150,000. With a renewed sense of urgency, Friends of the Land redoubled their fundraising efforts. They made an offer for $145,000 on January 23 and guaranteed a 10 percent down payment, but the executors wouldn't close until they had the full purchase price in hand. Meanwhile, other potential buyers began to express interest in the farm. Richland County's state representative, Neil Robinson, proposed that the state of Ohio purchase Malabar Farm as a site for a new state university or agricultural experiment station—or even as a cheap location for the "proposed north-south freeway," I-71. More troubling to Friends of the Land, a couple private groups considered turning Malabar into a country club or, worst of all, cutting the farm up into lots for a housing development.[21]

In February, another threat to Malabar arose as a logging crew moved into a couple trailers and a hut next to Bromfield's vegetable stand and set up a small sawmill in the Malabar woods. While hospitalized in New York shortly before his death, Louis Bromfield had signed a contract with the A. W. Hinchcliff Hardwood Lumber Company of Strongsville, Ohio, to harvest all trees over twelve inches in diameter on about sixty acres of the Malabar woods. What concerned Friends of the Land about the logging operation was not that trees were being cut. They were in favor of sound woodlot management, which included selective harvesting. The problem was how it was being done. Ever since 1946, two areas in the Malabar woods totaling thirty acres—a stand of old growth red oak, beech, and maple in the sugar bush and a nearby stand of second growth trees—had been part of the Ohio Agricultural Experiment Station's experimental forestry program. Foresters measured the trees every five years and designated which ones should be selectively harvested to give the others room to grow. Though Bromfield was in favor of scientific forestry in principle, he never actually cut any trees at Malabar during his lifetime. The contract with Hinchcliff, which Bromfield negotiated with the help of his friend Bill Solomon, went against the recommendations of OSU forester Ollie Diller, who wanted the trees to first be marked by a trained forester for selective logging. All the trees in the experimental forestry program were slated for the ax, and the actual logging operation was being conducted "most carelessly and destroying much timber."[22]

*Above:* Logging in the Malabar woods, winter 1957. (A/V box, from box 13, Friends of the Land Records Papers, courtesy of Ohio History Connection)

*Right:* Sawmill in the Malabar woods, winter 1957. (A/V box, from box 13, Friends of the Land Records Papers, courtesy of Ohio History Connection)

Unable to get an outright grant, Friends of the Land's leaders decided to try to borrow the money to purchase Malabar before it was too late, figuring they would be able to launch a more successful fundraising campaign if they had enough time. Ralph Cobey got together a group of Mansfield businessmen, called the Malabar Farm Trustees, which agreed to loan Friends of the Land $55,000 toward purchase of the farm. But that wasn't enough to meet the $145,000 purchase price, and on April 26, the executors of the Bromfield estate emphasized that "Malabar is still without a buyer." Assuming that Friends of the Land would never be able to come up with the cash, they started to draw up the paperwork to sell Malabar to someone who actually had the money to purchase it. Things looked bleak until, at what seemed to be the last minute, an idea occurred to Ollie Fink: Why not ask the Samuel Roberts Noble Foundation, in Ardmore, Oklahoma, if it would be willing to loan Friends of the Land the money to save Malabar?[23]

Wealthy oilman Lloyd Noble (1896–1950) established the Noble Foundation on September 19, 1945, in memory of his father, Samuel Roberts Noble. Lloyd Noble started his first oil-drilling business in 1921 and, like many others, made a fortune. Unlike many of his fellow industry executives, Noble often thought about the future. He knew that the oil would be gone someday but that the land would remain forever—that is, if it was taken care of properly. As he flew across the Oklahoma countryside in his private airplane from one oil field to another, Noble was sad to see

*From left:* Lloyd Noble, Louis Bromfield, and Thomas A. McCoy at the Noble Foundation's laboratory, circa 1949. (box 61, Friends of the Land Records Papers, courtesy of Ohio History Connection)

bare, eroded landscapes that had once been green prairies and productive fields. He became interested in soil conservation, made friends with Louis Bromfield, and served as a director of Friends of the Land until his death on February 14, 1950. Louis Bromfield served on an agricultural advisory board for the Noble Foundation in 1949, and his farm manager Bob Huge worked briefly for the Noble Foundation during the era of the ill-fated Texas Malabar project. The Noble Foundation even collected and analyzed soil samples at the Texas Malabar Farm in the summer of 1949. In fact, the Noble Foundation probably had a closer connection to Malabar Farm and Louis Bromfield than any other foundation in the country.[24]

Ollie Fink and Jonathan Forman happened to be traveling to Clinton, Oklahoma, on April 28–30 for the 1957 Friends of the Land Small Watershed Management Forum. Without an appointment, Fink took a taxi to the Noble Foundation's Ardmore headquarters and arrived just as Cecil Forbes, one of the foundation's leaders, was "leaving his office for a trip to the West Coast." Forbes, who had been a good friend of Louis Bromfield, listened attentively as Fink explained that Friends of the Land could purchase Malabar Farm for $140,000, but only if they had cash to make the purchase. The executors of the estate, he warned, were already in process of drawing up a sales contract with another buyer—which was rumored to be a real estate firm that planned to chop Malabar up into a housing development. "I trust you will appreciate the urgency of our situation and understand that it is not a case of trying to high pressure action on your part," he explained. After returning to Ohio, Fink sent Forbes an urgent telegram on May 6, informing him that a realtor from Mansfield planned to sign a sales agreement for the farm the following day. Only the Noble Foundation could save Malabar. But would it?[25]

To the relief and joy of Malabar fans across the country, the Noble Foundation agreed with Friends of the Land that saving Malabar was a worthy cause. At almost literally the last minute, on May 6, 1957, it sent a brief but monumental telegram to the Bromfield estate's executors: "This will confirm that Noble Foundation has agreed to lend $85,000 to Friends of the Land for preservation of Malabar Farm." It put the check in the mail the same day, and when the money arrived on May 8, Friends of the Land signed the paperwork to officially purchase Malabar Farm for $140,000, secured by a $55,000 mortgage from the Malabar Farm Trustees and $85,000 from the Noble Foundation. The telegram from the Noble Foundation had arrived just in time—"actually 30 minutes later and the place would have become a new Suburbia and only a memory in agriculture."[26]

"The heartwarming news from Mansfield, Ohio is that Malabar Farm, one of the world's most famous agricultural showplaces, is to continue as an experimental and educational center," Inez Robb announced in her nationally syndicated newspaper column. A ceremony transferring the farm's title from Bromfield's estate to Friends of the Land took place at noon on May 10 in the Big House dining room. It seemed like the

*Left:* Ceremony transferring the Malabar deed to Friends of the Land, May 10, 1957. *From left:* Herb Cobey, Cecil Forbes, Sam Noble, Ralph Cobey. (box 71, folder 1503, Louis Bromfield Collection, Ohio State University, courtesy of Ohio Department of Natural Resources)

*Below:* Ollie Fink (*left*) and Bill Solomon (*right*) holding the check from the Noble Foundation. (box 71, folder 1503, Louis Bromfield Collection, Ohio State University, courtesy of Ohio Department of Natural Resources)

generosity of the Noble Foundation might even open the long-awaited floodgate of funds. Bromfield's friend Doris Duke donated $30,000 to Friends of the Land, and actor James Cagney chipped in $1,000. Even the real estate agent from Mansfield who had been about to close on the farm, George Rogers, surprised everyone by donating $5,000 toward the ecological center. Rogers explained that he had planned to turn Malabar into "a riding academy, dude ranch and a golf course" and not a housing development but had "voluntarily dropped out of the picture when it developed that the Friends of the Land could raise the money."[27]

The first priority on Friends of the Land's agenda after receiving title to Malabar was to halt the destructive logging in the Malabar woods. At a meeting in Mansfield on May 15, 1957, the Malabar Farm Trustees and the leaders of Friends of the Land agreed "that timber cutting on the farm must stop." To reach the lumber company's headquarters as quickly as possible, Ollie Fink and Ralph Cobey chartered a plane and flew fifty miles to Strongsville, where they spent the afternoon negotiating an agreement to buy back the trees from the Hinchcliff Lumber Company. Desperate to stop the logging at any cost, they spent $11,500 to repurchase the timber that the lumber company had paid Bromfield $10,000 for—despite the fact that fifteen acres had already been logged. But at least they could say that the Malabar woods had been saved, and they decided to name a portion of the forest the "Doris Duke Woods" because the money to buy back the timber came out of her donation.[28]

Friends of the Land named the forest at Malabar "Doris Duke Woods" in 1957 or 1958 because the money to buy back the timber came from her $30,000 check. (A/V box, from box 1, Friends of the Land Records Papers, courtesy of Ohio History Connection)

Malabar was saved, at least for a year, when the mortgage from the Malabar Farm Trustees was due. But that was a long way off, and Friends of the Land hoped that, long before that time came, they would be able to raise $800,000 and start turning Malabar into an ecological center. Surely, they reasoned, Louis Bromfield had many friends who just wanted to say "thank you" for all his hospitality—and in this hope, they launched a national fundraising campaign targeting corporations, foundations, organizations, and individuals. Ollie Fink planned to publicize the project using a direct mail campaign, newspaper and magazine articles, and radio and television programs. To start raising public awareness about Malabar, he asked Paul Sears to write an article to send to *Reader's Digest* about the farm's history and the proposed ecological center. Sears was less confident than Fink that the campaign would be successful, but he wrote a nice article and tried to help as much as possible. Meanwhile, Friends of the Land, confident that money would soon come pouring in from hundreds of loyal Bromfield fans, started working on cleaning up the farm and getting it ready for visitors to tour once again.[29]

## Chapter 8

# Marginalization

Even though the influence of Friends of the Land was waning by 1957, the organization still had some friends in high places—even in the White House. Dwight D. Eisenhower had been on its board of directors of before he became president, and Ollie Fink sent him a letter to see if he would be interested in helping Friends of the Land gain publicity for fundraising efforts. While he didn't want to "lend his name in direct efforts for raising money for any cause," the president was sympathetic to the Malabar project and invited several leaders of Friends of the Land, including Ollie Fink and Jonathan Forman, to visit him at the White House. Newspapers printed articles and pictures of Friends of the Land members meeting with Eisenhower and reported that he "expressed pleasure" on hearing that they had saved Malabar and planned to operate it according to Bromfield's ideals. "I congratulate you on all you are doing in agricultural research, and in your efforts to promote conservation practices and to bring about an improvement of the productive soil in Ohio," the president told Ollie Fink.[1]

On August 4, 1957, Friends of the Land announced that tractor-drawn wagon tours of Malabar Farm would be offered each Sunday afternoon. The tours were led by W. Hughes Barnes, a biology professor at Muskingum College, who was an "outstanding field naturalist" and "well qualified to answer the questions of those interested in wildlife and other ecological relationships." In true Bromfield tradition, Friends of the Land didn't charge for the farm tours, though they did begin charging a "modest fee" for never-before-offered tours of the inside of the Big House. Along with the weekly Sunday tours, Friends of the Land launched a series of promotional events to get people in the local area excited about Malabar. At one event in September, they served tea to

Along with other equipment, Ralph Cobey donated this tractor-drawn tour wagon to Malabar. (A/V box, from box 1, Friends of the Land Records Papers, courtesy of Ohio History Connection)

two hundred prominent Mansfield citizens. A couple weeks later, they held an open house at Malabar in conjunction with their seventeenth annual meeting. The keynote speaker at the event was US senator Frank Lausche, former Ohio governor and friend of Bromfield, who exhorted Friends of the Land to continue Bromfield's legacy at Malabar. The Galion High School Band furnished music, while "seemingly endless files of visitors" toured the Big House and enjoyed "coffee and a light snack."[2]

While Malabar Farm received a lot of favorable local publicity, the national fundraising campaign was having some disappointing results. Shortly after Friends of the Land purchased the farm, Ollie Fink sent publicity information about the proposed ecological center to major magazines like *Fortune, Nation, McCall's, American Legion Magazine,* and *Parade.* Most of them politely told him they didn't want to write about Malabar, either because they didn't think their readers would be interested or because they had a general policy not to publish articles soliciting funds for any organization. Even the better-written article by Paul Sears fared poorly; it was rejected by *Reader's Digest, Life, Atlantic Monthly,* and *Harper's.* "We've managed to get ourselves so identified with conservation matters that we have to lean over backwards in the other direction in order not to wear out our welcome with our readers," the editor of *Harper's* told Fink. Even Cass Canfield, the editor at Harper & Brothers who had published Bromfield's books, was hesitant

Brochure for Louis Bromfield Institute, 1958. (box 139, folder 2263, Louis Bromfield Collection, Ohio State University, courtesy of Ohio Department of Natural Resources)

to endorse the fundraising campaign, because he wasn't sure the donors would get their money back if insufficient funds were raised to complete the project.[3]

All in all, the initial fundraising campaign fell far short of Ollie Fink's anticipations. By September 1957, Friends of the Land had solicited approximately $45,000 for the Malabar project, including the $30,000 check from Doris Duke. None of this money was used to pay off the mortgage; it went to save the Malabar woods, repaint the Big House, conduct needed repairs, and mow the lawn. Still convinced that thousands of people would want to contribute to the Malabar cause if they could be properly persuaded, Friends of the Land decided to hire a professional fundraising agent from Cleveland, Robert L. Beda, for a fee of $1,000 a month. Ever optimistic, Friends of the Land assumed Beda's fundraising efforts would bring in more than enough money to pay him. They also enlisted Louis B. Seltzer, a newspaper editor from Cleveland, to help with the publicity angle and form a Committee of 100, a group of influential people who would lend their support and prestige to the fundraising effort. Seltzer explained to potential committee members, "Lending your name to this committee will not obligate your time, but rather [be] an endorsement which will strengthen our appeal for funds to finance the program."[4]

The renewed fundraising campaign was launched in December 1957 with a wave of newspaper publicity and a professionally printed blue and gray brochure announcing plans for "The Louis Bromfield Institute." The brochure's cover had a photograph of the Big House next to a conceptual drawing of a huge institutional building, proudly flying the American flag. The Louis Bromfield Institute, Friends of the Land proclaimed, would "be a symbol to the world of America's concern for the welfare of humanity." "Conservation of our water, land and natural resources and the meeting of the problems of food production, nutrition and the harnessing of nature to aid and benefit mankind is not only America's number one problem but a world-wide concern," they argued. "The Louis Bromfield Institute is a step toward the solution of this basic problem. The task before it is vital and necessary. It must succeed." Endorsing the Louis Bromfield Institute were many longtime conservation leaders like Bryce Browning, Morris L. Cooke, and Hugh Bennett, along with celebrities like James Cagney and Doris Duke. Just as notable as who was on the list was who was *not* on it—on the one hand, no soil researchers or agronomists except William Albrecht were on the Committee of 100, but on the other hand Forman specifically told Beda not to invite J. I. Rodale to join the project.[5]

Even before the Louis Bromfield Institute became a reality, it hired its first employee—a wildlife ecologist named Floyd B. Chapman (1911–1984). Chapman earned his PhD in zoology from the Ohio State University in 1938; his bachelor's and master's degrees were in botany. He worked as a game manager for the US Fish and Wildlife Service for two years and then spent seventeen years in the Ohio State Division of Wildlife. By January 1958, Chapman was growing frustrated with the limitations of his government job. He wanted to do basic research in wildlife ecology but felt like his research ambitions had always been "thwarted" by "being pressed into administrative positions, where I have long felt that my training was being wasted in routine approval of purchase orders and invoices." On January 7, 1958, Chapman wrote to Paul Sears that he had "a very strong desire" to "become a member of the staff of the Bromfield Institute." After Chapman resigned from his government job, Ollie Fink asked him if he wanted a job doing "library work and other tasks in setting up the Information Center." Chapman jumped at the chance, Friends of the Land decided that "there was no one in Ohio better qualified to do this job for us," and they hired him at a salary of $5,000 per year with the job title of acting director.[6]

One of the first things Chapman did as acting director was arrange a colloquium on "the ecologic pattern of Malabar Farm," held on April

Visiting naturalists exploring the Jungle at Malabar Farm, 1958. (A/V box, from box 1, Friends of the Land Records Papers, courtesy of Ohio History Connection)

26 and 27, 1958. Twenty scientists from various Ohio universities and government agencies presented a "gold mine of information" about every aspect of Malabar's ecology—geology, soils, vegetation, wildlife, water, forests, farming methods, and economics. It took two entire issues of *Land and Water* to print the conference proceedings, which Friends of the Land intended to use as a baseline for future research projects at Malabar. In mid-May, Chapman invited a "group of naturalists from various sections of the state" to help plan a nature trail in the Malabar woods. After walking around the farm, they concluded that Malabar was a perfect place for a nature trail, with diverse forests and streams, sandstone outcroppings, caves, and rare plants. The first mile of nature trail, which started at a "green gate by the Big House," opened to the public on September 1, 1958. Highlights on this trail included the now unnecessary diversion ditch called "Hecker's Folly"; the old orchard and beehives; "The Lonesome Road," an abandoned township road; an outcropping of Blackhand sandstone, "the Buzzard Rocks"; and mosses and ferns on a vertical rock face called the "Hanging Gardens."[7]

As visitors started to tour Malabar again, Friends of the Land worked on getting the farm's agricultural operations back in shape. They reopened the vegetable stand on summer Sundays in 1958, but their main emphasis, based on Bromfield's glowing reports of its profitability in the late 1940s, was to revitalize the Malabar dairy operation. In March 1958, they had an agricultural consultant from Illinois named Pete Cooley inspect the farm and give them advice. While acknowledging that Malabar was a good location for dairy farming, Cooley warned Friends of the Land, "It can and will lose you lots of money without excellent operators and very high producing cows." Based on his own experiences with dairy farming, Cooley warned them "to go slowly—slowly—slowly—before putting all your eggs in one basket." This was sound advice and very similar to what Louis Bromfield had said when he downsized his dairy herd in 1954. Half of the dairy farms in Richland County stopped producing milk between 1949 and 1959, and both the amount of milk produced per farm and the amount of capital required to be successful had greatly increased since the golden age of grass farming in the late 1940s. As the Ohio Agricultural Experiment Station said in 1959, "If any one word typifies the Ohio dairy industry in the past ten years, it is *CHANGE.*"[8]

Friends of the Land did not heed Cooley's advice. They pushed ahead with expanding the dairy operation, which was down to thirty-five lactating and dry cows and eighteen heifers, back to where it had been around 1950. One of the first things they had to do was spend $4,500 to buy a five-hundred-gallon bulk milk tank and compressor. Bulk milk handling, first used in California mega-dairies in 1938, came to most Ohio milk markets in the mid- to late 1950s. In a bulk handling system, the milkman arrived at the farm in a tanker truck and pumped milk directly out of a farm's bulk tank, which had many advantages over the old milk can system. It cooled the milk faster, was more sanitary, reduced the cost of handling, eliminated the heavy labor of lifting milk cans, and allowed each truck to haul more milk. The only major disadvantage was that bulk tanks were expensive, but farmers had to purchase them if they wanted to continue to sell their milk. Malabar had to buy a bulk tank when the Borden Company switched its Mansfield processing plant over to bulk milk handling in October 1958. Considering that their annual milk check was only $14,000, the new bulk tank was a sizeable investment for Friends of the Land. Yet they still believed that "the farm will contribute to paying off the debt incurred by the purchase and in the future provide funds for expanding areas of research."[9]

Despite continued positive publicity and grandiose plans for the dairy farm, Friends of the Land's Malabar project was in serious financial trouble. The anniversary of the farm's purchase came and went without a single dollar being paid on the mortgage. As the summer of 1958 crept on, it became painfully obvious that the fundraising campaign headed by Robert Beda was worse than a failure—it actually cost Friends of the Land $6,400 more to pay Beda than he raised. Meanwhile, the Malabar Farm Trustees were politely but firmly demanding their money back. When Ollie Fink explained that Friends of the Land couldn't pay, the Mansfield businessmen threatened to foreclose. To make things worse, Friends of the Land had put so much time and money into the Malabar project that they hadn't been able to recruit many new members. The organization had 2574 members in 1956 and only 2023 in 1958. At a board of directors meeting in September, Forman suggested that the best solution to both problems would be to make the Louis Bromfield Institute a separate organization. Of course, mere reorganization wouldn't pay off the mortgage—Friends of the Land needed money, and fast.[10]

It was in the midst of this crisis that Friends of the Land elected Ralph Cobey president. An industrialist from Galion, Ohio, Cobey had a degree in mechanical engineering from Carnegie Mellon University and joined his father's manufacturing company, the Perfection Steel Body Company, in 1930. Ralph Cobey was president of Perfection from 1945 to 1970, acquired the Eagle Crusher Company, and established the Cobey Corporation to make farm machinery, including one of the first power-driven manure spreaders. In addition to his manufacturing businesses, Cobey was interested in agriculture and gardening and was well known in the Galion area for his philanthropic works. Cobey's connection with Louis Bromfield had been a business one: like many other manufacturers, he loaned Bromfield lots of farm machinery to demonstrate at Malabar. His brother Herbert was Bromfield's friend, and the Cobey brothers oversaw the farming operations at Malabar during the uncertain period between Bromfield's death and Friends of the Land's purchase of the farm.[11]

After Bromfield died, Massey-Harris-Ferguson arrived at Malabar Farm with a trailer and took away its tractors; many other manufacturers followed suit and removed their machinery from the farm. To fill in the gap and keep the farming operations going, the Cobeys loaned even more machinery to Malabar. Ralph Cobey also formed the Malabar Farm Trustees to help save Malabar and loaned $15,000 to the cause—more than anyone else in the group. When, in the summer and fall of 1958, the other businessmen started demanding their money back, Cobey paid them off and assumed their part of the mortgage to give Friends of the Land more time to raise money. When he became president of the group in September and realized the dire financial situation of the Malabar project, Cobey took some drastic action. He appointed an executive committee, of which he was in charge, which authorized him to make the decision to form a new foundation and transfer title of Malabar away from Friends of the Land. With every other fundraising attempt ending in failure, Friends of the Land turned to the only place left—the Noble Foundation.[12]

In early November, Ralph and Herb Cobey, their wives, and Bromfield's old friend and neighbor Bill Solomon flew to Ardmore, Oklahoma, to meet personally with Sam Noble and Cecil Forbes. After a couple "strenuous" days of touring oil drilling operations and ranches, at a big party their last night in Ardmore Solomon finally got up the nerve to be honest about the financial situation. He asked if the Noble Foundation could take over Malabar Farm. Forbes explained that wouldn't be possible; instead he suggested that the Noble Foundation take on the entire mortgage, pay off the Mansfield businessmen, and form a new foundation called the "Louis Bromfield Malabar Farm Foundation." "I was exhausted mentally but will never be any happier," Solomon wrote later. "Herb and Ralph sat there with their mouths hanging open, flabbergasted." After some discussion about logistics, the Louis Bromfield Malabar Farm Foundation was incorporated in Columbus as a separate nonprofit organization on December 17, 1958, with the sole purpose of conducting "a conservation, education and research program in agriculture" at Malabar Farm. Friends of the Land transferred the title and mortgage on Malabar to the new organization, and the Noble Foundation paid off the Mansfield businessmen and assumed the entire mortgage—except for the $15,000 Ralph Cobey had loaned. The Noble Foundation thought it best for Cobey to maintain a financial interest in Malabar's success, though initially he asked them to buy him out along with the rest of the Mansfield group.[13]

While sad that Friends of the Land had to give up the title to Malabar and a little frustrated that Ralph Cobey was bypassing normal procedures, Ollie Fink was initially pleased with the move to create the new foundation. By the end of December, however, he was very concerned about what he perceived as Cobey's attempt to take over Malabar Farm for his own interests. Cobey had arranged to have his private attorney handle the legal side of the transfer, and when the initial three trustees of the new foundation were announced, they were Ralph Cobey; Jonathan Forman; and William Locke, an insurance agent and former mayor of Mansfield who had been one of the Malabar Farm Trustees. Ollie Fink believed Cobey had purposely excluded him from leadership of the Malabar project and suspected that Cobey had joined Friends of the Land solely for the purpose of taking over Malabar so he could turn it into a "show case exhibition of Cobey farm equipment." Cobey appointed one of his sales representatives, Charles Clark, as farm manager and incorporated a sales pitch for Cobey machinery into the farm tours. And Cobey wasn't committed to farming exactly the way Louis Bromfield did. "I didn't know Mr. 'B' like you did but you can't forget he is dead," he told Bill Solomon.[14]

Jonathan Forman tried very hard to stay neutral in the controversy and to reconcile Fink and Cobey. "I am writing to try and straighten you out before, in confusion, you do something more and destroy all the work you have put into Friends of the Land through the last 18 years and all your good work for the Malabar project," Forman wrote to Fink. While acknowledging that Cobey did want to use Malabar to demonstrate his machinery, Forman urged Fink to look at Cobey's philanthropic background, the fact that he had "put his soul" into the effort to save Malabar, and that he had done "a good job in the operation of the farm." The real problem, Forman explained, was "trying to swing $150,000 without funds." Forman even suggested that Fink's angry letters about Ralph Cobey to the Noble Foundation were one reason that he was not considered to be one of the trustees. "You and I are old friends," Forman concluded. "We have worked at this too long to see all of it go down the drain."[15]

Even after reading Forman's letter, Fink remained convinced that the situation was "wrong and dishonorable." "I am about fed up with compromising with underhand dealing," he said. "I would much rather sever my connections with FOL and use the rest of my life publicizing the dirty deal we have had at Malabar." Fink was so convinced that he had a "moral obligation" to stand up for what was right that he was unwilling to talk or correspond with Ralph Cobey personally after the December 17 meeting, or to apologize for anything until Cobey apologized to him first. When Bryce Browning heard Fink was planning to publicize his allegations against Cobey, he was greatly disturbed. "While fully appreciating your position and the resulting feeling that action must be taken, I fear that the taking of the proposed action at this time would mark the beginning of the end, so far as Friends of the Land is concerned," he warned. "It would seem that the controversy resulting from your proposed action could only result in a 'house divided against itself.' In such event Friends of the Land might be destroyed and the Bromfield Memorial program also dealt a death blow."[16]

"I would be greatly disappointed if I should be responsible for any disaster to Friends of the Land," Fink replied. But he confided to Browning that the organization was already in dire financial straits, especially since all the money it put into the Malabar project had been deeded over to the new foundation along with the farm. With all the emphasis on Malabar, Fink hadn't had time or funds to recruit new members, and membership was dropping below two thousand. Since Friends of the Land still charged the same membership fee they had in 1940, there wasn't enough money to pay Fink's salary—let along hold four or five annual conferences and publish a magazine. Two issues of *Land and Water* were put out in 1959, but that was all. There was no money for more. And this time there were no wealthy conservationists to bail out Friends of the Land, no influential figures to crusade around the country—and, sadly, no public groundswell for soil conservation. Reluctantly, Fink and the board of directors concluded that the only solution was to merge Friends of the Land with a more fiscally stable conservation organization. Early in 1959, Fink entered into negotiations for a possible merger with the American Forestry Association, but the two groups were unable to reach a mutually acceptable agreement, and the deal fell through. Toward the end of the year, Fink decided to try again and merge the nearly defunct Friends of the Land with the Izaak Walton League of America.[17]

The Izaak Walton League, "Defenders of Soil, Woods, Waters and Wildlife," was an older and larger conservation organization. Originally established in 1922 by a group of fifty-four outdoorsmen "to stop the widespread and arrogant abuse of America's rich heritage of fish, wildlife and scenic resources," the league had about 50,000 members in 1959 and published a short monthly magazine called *Outdoor America*. Unlike Friends of the Land, the Izaak Walton League emphasized direct political action and was especially active in promoting and encouraging the preservation of wilderness areas. In 1959 it launched an "SOS" campaign—"Save Our Shorelines"—to support the proposed addition of several notable shoreline areas to the national park system and to preserve smaller but locally important shorelines all across the country. The league was also influential in helping design the federal Soil Bank program, which paid farmers to retire marginal acreage from production and instead use it for wildlife habitat or recreation. Other issues it addressed in the late 1950s and early 1960s included water and air pollution, littering, destruction of wildlife habitat, and indiscriminate pesticide use.[18]

Frank Gregg, the president of the Izaak Walton League, was initially skeptical at the proposal to merge Friends of the Land with the league. Yes, both organizations were interested in conservation—but he saw Friends of the Land as "a highly specialized group in the sense of a special concern for what one might call natural farming and land use methods—something like the organic gardening philosophy, only expanded to apply it to the entire problem of land use." He thought members of Friends of the Land might be "sorely disappointed" that the Izaak Walton League, while sympathetic to the cause, did not focus much on soil conservation. After some discussion, however, the league decided that the two organizations both had very broad purpose clauses and did emphasize the same issues, especially when it came to water conservation. Though there was some talk about trying to perpetuate the name of Friends of the Land, it soon became clear that the proposed "merger" would really mean that Friends of the Land would be dissolved and its assets and membership absorbed into the Izaak Walton League. By February 1960, negotiations had proceeded far enough that Ollie Fink mailed out a letter to the members of Friends of the Land asking their opinion on the proposal.[19]

Reactions were mixed. Many agreed that Friends of the Land had been declining for several years and thought merging with the Izaak Walton League was the best option available. "I had just about decided to drop membership in 'Friends,' after being a member since the organization started nearly twenty years ago," wrote David W. Dresbeck, a soil conservationist from California. "I'm sure that you, Ollie, have been well aware of the decline in its vitality as a stimulating influence in conservation. This is understandable, because all human efforts depend on people to make them succeed, and 'Friends' has gradually suffered from the loss of some of its strongest leaders." "It is, indeed, sad and distressing there are to be no more Friends of the Land," a forestry consultant from Ohio lamented. But he and several other members thought the Izaak Walton League was a "wise choice" if Friends of the Land had to merge with another organization, especially "now that Malabar Farm has been set up as a separate entity." Others were opposed to the merger and did not think that the League was a good fit; six or eight of the letters received were "hostile." But "the majority clearly favors the consolidation," Frank Gregg reported to Ollie Fink, concluding, "This appears to be plenty of encouragement to go ahead substantially as outlined."[20]

On March 24, 1960, Ollie Fink and Frank Gregg agreed to merge Friends of the Land with the Izaak Walton League. Friends of the Land was officially dissolved, a complicated legal process that was not completed until February 20, 1961. A short news item announcing the consolidation appeared in the April 1961 issue of the Izaak Walton League's *Outdoor America,* and a two-page article about Friends of the Land was printed in the May 1961 issue. From April 1961 until November 1961, the *Outdoor America* masthead contained the words, in very small type, "Consolidated with 'Land and Water.'" Until January 1963, the renamed *Izaak Walton League Magazine* included a note under the table of contents that Friends of the Land had been consolidated with the Izaak Walton League. And that was all. The magazine's content did not change to reflect the "broader" program that some leaders of Friends of the Land had envisioned. The Izaak Walton League was still the Izaak Walton League, and Friends of the Land was no more.[21]

The Izaak Walton League had some difficulty reaching a settlement with Ollie Fink, who wanted the league to pay him and his wife, Julia, salaries and rent for storing Friends of the Land's office furniture during the merger and dissolution process. After a lot of negotiation, the league took about half of the furniture and left the rest and an electric typewriter with Fink to keep him from charging more rent. It also left all of the documents Friends of the Land had accumulated over twenty years, which Fink stored in his barn and planned to sort through at some later date. Fortunately for historians, he never got around to it, and Julia Fink and Jonathan Forman donated the papers, somewhat mouse-eaten and water-damaged but mostly intact, to the Ohio Historical Society after his death. Meanwhile, Ollie Fink had gone back to work as a schoolteacher at the Grover Cleveland Junior High School in Zanesville in 1959. With his previous years of teaching and working for the State of Ohio, he only needed to work four more years to receive a greatly increased retirement pension. He retired in 1963 and continued to be interested in conservation and history until his death in 1970.[22]

Perhaps the saddest part about the formation of the Louis Bromfield Malabar Farm Foundation and the subsequent dissolution of Friends of the Land was the infighting between Ollie Fink and Ralph Cobey. There is no record of whether the two men ever reconciled, but since Fink was still trying to get the farm away from Cobey in 1962, they probably never did. It was, as Russell Lord aptly observed, "a sad finale to a citizens society starting 20 years ago 'For the Conservation of Soil, Rain and Man,' and ending in frantic efforts to conserve itself." Sadly, the whole argument was based on misunderstandings. Cobey was unfamiliar with the history of Friends of the Land and saw it as a dying organization that wasn't worth saving; Fink was still trying to live in the past and was reluctant to accept the reality that the world had changed dramatically since 1940.[23]

Although the effort to save Malabar might have been the last straw that bankrupted Friends of the Land, it is doubtful that the organization would have lasted much longer regardless. One problem was that the organization's leaders and membership were aging. "I don't see any younger men coming on to take our place and without younger men we are headed exactly nowhere," Louis Bromfield warned back in 1952. Friends of the Land was never financially stable, and the declining quality of their publications during the 1950s certainly didn't help recruit new members. But the underlying cause of the organization's decline was much deeper: American society was undergoing a paradigm shift that marginalized everything Friends of the Land, Louis Bromfield, and Malabar Farm stood for.[24]

The reality was that Louis Bromfield died and Malabar Farm went up for sale at one of the lowest points for the conservation/environmental movement in the entire twentieth century—the late 1950s. With Cold War tensions heightening, national security took precedence over conservation. Conservationists like Hugh Bennett and Paul Sears watched with dismay as, beginning in 1958, the nation's fixation on the space program superseded seemingly mundane concerns about the wise use and protection of natural resources. Physics, chemistry, and engineering were exalted as the most important sciences, while the biological and agricultural sciences were pushed to the back burner. To make things even more difficult, the media labeled as "cranks" anyone who challenged this lopsided emphasis on chemistry and technology—like J. I. Rodale and his organic gardening movement. Scientists who dared question the new paradigm faced the very real possibility of losing their positions and credibility, and so most stayed silent, even if they were personally concerned about the direction things were taking. So real and intense was this censorship—though so subtle that the public was largely unaware of it—that one historian called the late 1950s "the McCarthy Era of the Conservation Movement." Those who tried to stand up against the tide paid the penalty—and one of them was Jonathan Forman, whose antifluoridation views cost him a prestigious scientific position.[25]

In the early 1950s, public health officials began promoting municipal water fluoridation as a preventative measure against tooth decay. But antifluoridationists argued that sufficient testing had not been done, it was impossible to control the dose because everyone drank different amounts of water, and too much fluoride could have detrimental effects on human health. When the City of Columbus first proposed municipal water fluoridation in 1956, Forman took an antifluoridation stance in several widely publicized debates, arguing that fluoridation was "mass compulsory medication." Thanks in part to his efforts, two-thirds of Columbus citizens voted against a fluoridation referendum in May 1958. By this time, however, the Ohio State Medical Association had officially endorsed fluoridation, and it wanted unanimous agreement from its officials and employees. The association gave Forman an ultimatum: either he could stop publicly fighting against fluoridation or he could resign from his position as editor of the *Ohio State Medical Journal.* "I took the stand that my outside activities were my own business," Forman said, "and that my work against fluoridation is more important to the American people than the editorship." After twenty-two years, Forman lost his editorial position, but his efforts helped keep Columbus's water unfluoridated until 1973.[26]

Whether the issue was fluoridation, organic farming, or pesticides, anyone who disagreed with the party line of scientific consensus in the 1950s was discredited. The middle-of-the-road stance Louis Bromfield advocated was no longer a viable option, which was one reason that Friends of the Land could not make Malabar the ecological center of their dreams. Even the scientific discipline of ecology itself was changing. Most of the land-use planning of the 1930s and 1940s had been based on the plant ecology of Frederic Clements, who believed that there was an inherent order and balance in nature and that an ecological community was like a complex organism, greater than the sum of its parts. In the 1950s, however, many ecologists discarded Clements's theories in favor of the "individualistic concept of the plant association" advocated by botanist Henry Gleason. Gleason argued that there was no order or balance in nature; the assortment of vegetation in a given region was purely a "coincidence." A wave of anti-Clementsianism fragmented ecology in the 1950s and served to weaken the very movement toward ecological agriculture that Friends of the Land was trying to promote. While a form of the holistic ecology on which Friends of the Land had been founded did continue in the subdiscipline of ecosystem ecology, ecologists in the 1950s and 1960s focused almost exclusively on studying natural ecosystems. The application of the ecosystem concept to managed systems in agriculture did not gain much acceptance until the 1970s.[27]

Friends of the Land, like Louis Bromfield, correctly saw that the application of ecology to agriculture would be essential to build a permanently productive agricultural system. But they were completely unprepared for the unprecedented hostility to their ideas that emerged in the late 1950s. Far from becoming more ecological, both mainstream agriculture and official USDA policies consciously turned against the goal of working with nature that had been so prominent in the New Deal conservation movement. Battle lines were drawn between "scientific" chemical-based agriculture and the "pseudoscience" of organic farming advocated by J. I. Rodale. The idea that science and nature could work together to make a better agriculture, which Louis Bromfield had promoted and Malabar Farm symbolized, was never mentioned. It was as if the whole soil conservation movement had never happened, as if the only choices were inefficient pioneer farming (with which organic agriculture was inaccurately equated) and "modern" chemical agriculture. What is most revealing about agricultural histories written during the Cold War era is that the whole permanent agriculture and soil conservation movement was usually not even mentioned. As the old conservation leaders died or retired, the onward rush of technological optimism marginalized the cause to which they had devoted their lives.

By 1960, Malabar Farm was a symbol of something that many people wanted to pretend had never happened. Given the intense hostility to natural farming methods that was developing right when they were trying to save Malabar Farm, it is not surprising that Friends of the Land were unable to raise support for their project. The real miracle, all things considered, is that Malabar was saved at all.

## Chapter 9

# Surviving the Sixties

To the general public, the transfer of Malabar Farm from Friends of the Land to the Louis Bromfield Malabar Farm Foundation was a smooth transition; nothing significantly changed in the way the farm was managed. As before, visitors interacted mostly with Floyd Chapman, whom the foundation kept employed as its resident ecologist. Chapman lived in the Big House, led guided hikes on the farm's nature trails, and wrote an informative monthly *Malabar Farm Newsletter* about what was going on at the farm. Tourists from all over the world came to tour the Big House, view the original furniture arranged like the Bromfield family had just stepped out of the room, and listen to Malabar legends that grew with every telling. It seems to have been Jonathan Forman who started the tradition of telling farm visitors that the little Ceely Rose house on Bromfield Road was haunted. In Forman's version of the story, he told a new tenant, "The ghosts walk there every night, the old miller and his wife go thumping around and it is very disturbing." The tenant thanked him for the heads-up and said, "I shan't mention it to my children or to my wife but I'm going to give it in full detail to my mother-in-law."[1]

After their tour of the Big House, visitors could stop by a new souvenir shop in one of the small buildings across the driveway. Items for sale included photos of Bromfield, honey from the Malabar Farm beehives (which Forman managed), blue-and-white souvenir plates depicting Malabar scenes, postcards of the farm, Bromfield's farm books, and a select few titles by Paul Sears, William Albrecht, and other authors. The gift shop was Forman's brainchild; he started it in the fall of 1958, paid for the inventory out of his own pocket, and gave all the profits to the foundation. Visitors could also look at a demonstration beehive donated by the A. I. Root Company of Medina and get a glass of pasteurized, homogenized whole milk for ten cents from a refrigerated milk dispenser, leased from and refilled by the Borden Company. During the summer, they could stop by the roadside vegetable stand to purchase staple vegetables like sweet corn, cabbage, beets, cucumbers, green beans, pumpkins, and watercress, all grown without pesticides. Or they could try some of the novel crops Chapman was experimenting with, like "curious pod corn from South America" and the "tomango," which he described as "a hybrid of the tomato and mango pepper." Malabar-grown herbs like basil, thyme, savory, marjoram, rosemary, sage, anise, and dill were dried and sold in the gift shop, along with popcorn, pickles, and jellies.[2]

Visitors to Malabar in the early 1960s could take a tractor-drawn wagon tour of the farm on summer Sundays or drive their own cars up to the summit of Mount Jeez on a new road, constructed with gentler grades than the one Bromfield had used. Or they could take a guided hike on one of the Malabar nature trails with Chapman. The trail behind the Big House, named the Rimrocks Trail, was so popular that the foundation constructed a second nature trail in 1959. This one, in the swamp on the east side of the farm that Louis Bromfield had called the Jungle, featured water-loving plants such as skunk cabbage, scouring rushes, sycamores, and willows. Other points of interest on the Jungle Trail were Bromfield's swimming hole, natural gas wells that had been in production at Malabar since before Bromfield bought the property, and a pile of bleached bones called the "cow graveyard" where Bromfield had dumped the carcasses of deceased cattle. In August 1959, Chapman designed interpretive signs for the nature trails and opened them for self-guided hikes in addition to guided tours.[3]

Malabar was more than a mere tourist attraction; at least for a few years, the foundation did as much as it could to turn the farm into an ecological center, despite a limited budget. One of Chapman's first responsibilities was to get the ecology library up and running. He catalogued several thousand conservation and ecology-related books and journals according to the Library of Congress system, including books formerly owned by Bromfield, a thousand titles Forman donated from his own collection, and forty-five children's conservation books. The books were shelved in the large sandstone-walled room in the Big House basement where Louis Bromfield had once held a lobster party and Ellen had canned hundreds of jars of peach preserves. The foundation put maroon and cream linoleum squares on the concrete floor, lined the walls with bookshelves, and provided tables and chairs for reading. A large portrait of Hugh Bennett looked down on visitors to the library, and old Soil Conservation Service posters proclaimed "Soil is Sacred" and "A six inch layer of soil keeps you alive." On the wall above the fireplace hung a wreath made from seeds set in wax, collected in the Pleasant Valley area back in the 1880s.[4]

A second basement room, which had formerly housed Bromfield's pool table and other games, was remodeled into a meeting room, complete with chairs, tables, and slide and movie projectors. Floyd Chapman and Jonathan Forman used this room for lectures on diverse topics, including "Prehistoric Indians" and "Soil Fertility and Health." In February 1959, Friends of the Land held its last colloquium at Malabar, on the dangers of uncontrolled, unplanned suburban sprawl. They planned to hold their eighteenth annual soil and health conference at Malabar in June, but the event was canceled at the last minute, probably due to the sour relations between Ollie Fink and Ralph Cobey. Some people were already on their way and didn't find out that the conference had been canceled until they arrived, so Forman and Chapman "hastily put together an informal program of tours of the farm and nature trails and forays into the nearby Muskingum Conservancy District." Forman revived the soil and health conferences at Malabar in 1960 under the sponsorship of the Louis Bromfield Malabar Farm Foundation, and by 1962 attendance was back up to about two hundred people.[5]

Another goal of the foundation was to make Malabar into a model dairy farm to demonstrate that good farming was profitable. To get the most up-to-date scientific advice for their farming operation, Floyd Chapman organized an Agricultural Advisory Board, chaired by John H. Sitterley, an agricultural economics professor at the Ohio State University. Two other OSU professors—agricultural engineer Joseph D. Blickle and dairy scientist C. D. McGrew—also served on the board, along with a couple local farmers. At its initial meeting in January 1960, the advisory board explained that it would be unrealistic to assume that the farming operation would make enough money to cover anything except its own expenses. It emphasized that only a large, efficient, well-run dairy operation could stay in business in 1960, and so it developed a plan to double the dairy herd from forty to eighty milking cows and construct a new, modern dairy barn on the corner of Pleasant Valley and Bromfield roads. After discussing various designs, board members settled on a loose housing layout with two four-cow milking parlors, along with a special viewing area for visitors to watch the milking through a glass window. Once the new barn was completed, Chapman planned to turn the old dairy barn into a "little theater" that would seat about a hundred people.[6]

Under Chapman's excellent leadership, the foundation offered educational programs to groups of all sizes and ages. The most ambitious project was the Malabar Junior Explorers Program, a partnership with the Mansfield public school system to bring its entire fifth grade

Boy Scout camp on Mount Jeez, circa 1961. (Malabar Farm archives, courtesy of Ohio Department of Natural Resources)

class on a field trip to Malabar Farm. Nearly a thousand schoolchildren participated in the program in 1961 and 1962, arriving in groups of forty students twice a day for most of the month of May. Both students and teachers gave the program "enthusiastic applause," and one class of children even painted a mural about conservation and gave it to the foundation as a thank-you gift. To help bring young people closer to nature and agriculture, the foundation allowed groups of Boy Scouts, Campfire Girls, and other outdoors clubs to camp at Malabar, usually on Mount Jeez. And individuals and groups from all over the world continued to come to Malabar. One Swiss visitor hiked all the way from Lucas in the pouring rain, and the mayor from Mansfield's sister city in Tanganyika considered visiting Malabar the highlight of his trip to Ohio.[7]

In keeping with Bromfield's tradition of hosting international students at the farm, the foundation held a Louis Bromfield International School of Practical Agriculture during the summer of 1961. Three international college students, from Iran, Israel, and Mexico, spent the summer at Malabar. They lived in the Big House and went on field trips to local attractions, including the Ohio Agricultural Experiment Station in Wooster, the Coshocton Hydrologic Station, several farm machinery factories, and an "outstanding sheep farm." The climax of the summer was a visit from the Iranian ambassador, Adeshir Zahedi, who flew into the Mansfield airport and gave a speech of international peace and goodwill at Malabar on August 9. After the ambassador's speech, a crowd of guests enjoyed a picnic lunch on the lawn in front of the Big House, partaking of American staples like fried chicken, baked beans, and potato salad, along with some international dishes like hot tamales and tortillas. A group of twenty-eight international students came from OSU to hear the ambassador, but unfortunately their bus broke down and they didn't arrive at Malabar until after Zahedi had left.[8]

The International School of Agriculture was just the beginning, the foundation told visitors. It hoped to host ten or twelve students the following summer at a "Louis Bromfield Malabar Farm School of Applied Agriculture," where both international and American young men would "live and work at Malabar" and "soak up Bromfield's ideas as they soak up the ideas that helped produce them." It planned to build a permanent "Malabar Youth Camp" at the Ferguson Place, possibly named after former Mansfield mayor William J. Locke, who had been one of the trustees of the foundation until he passed away in July 1961. The foundation still hoped it would be able to make the ecological center a reality, if it could just manage to get enough money. The Noble Foundation did not expect Malabar to pay back its loan "for a considerable period of time," and both parties planned to renegotiate the mortgage in 1966, when the promissory notes came due. Since it was not expected to make payments on the mortgage right away, the Louis Bromfield Malabar Farm Foundation could put any money it raised back into the farm and programs.[9]

To help raise publicity, the Louis Bromfield Malabar Farm Foundation initially mailed out the *Malabar Farm Newsletter* free of charge to three thousand people, including most of the former members of Friends of the Land. Once it attained tax exempt status from the Internal Revenue Service at the end of 1959, the foundation encouraged newsletter subscribers to purchase memberships, starting at five dollars a year. A year later, only six hundred people had become foundation members, while the number of names on the newsletter mailing list had grown to five thousand. In April 1961, the foundation informed everyone on the mailing list that it was getting too expensive to send out so many free newsletters and that henceforth only members would receive the monthly newsletter. Meanwhile, its leaders tried every possible way they could think of to raise more money. The foundation briefly hired a public relations director named John Stark, who tried to capitalize on year-end giving in 1961. Chapman helped raise publicity for the farm by giving a series of lectures on Malabar at the Mansfield Public Library, timed to coincide with the release of Ellen Bromfield Geld's 1962 book about her memories of Louis Bromfield, *The Heritage*. But the fundraising campaigns were largely unsuccessful, and Stark's position was terminated in June 1962.[10]

With the failure of the fundraising campaign, plans for an expanded dairy herd and new barn had to be shelved. Instead, some remodeling was done on the old dairy barn and the foundation spent $2,500 to concrete the muddy outside feedlot. Since the old barn could only house a maximum of forty cows, the Agricultural Advisory Board decided at its final meeting, in March 1962, that the Malabar dairy herd couldn't become a cutting-edge modern farm without sufficient funding. Members suggested diversifying the farm operation again, keeping the dairy herd small enough to fit in the existing barn, and taking advantage of the federal Feed Grain Program of 1961 and 1962 to get payments for the decrease in corn production that they would be making anyway. It didn't matter that the Malabar dairy herd was the largest in Richland County, that its 1961 milk production of 486,890 pounds was four times the regional average, that the foundation had sold off most of Bromfield's old cows and replaced them with a herd of high-producing pedigree Holsteins, or that the Malabar cows had some of the highest milk production records in the county. Instead of proving that good farming paid off, the Malabar herd demonstrated an unpleasant reality—even good dairy farming that followed expert advice wasn't profitable in the 1960s.[11]

While milk production per cow increased steadily during the 1950s and 1960s, per-capita milk consumption declined, and the dairy industry was constantly faced with problems of surpluses and low prices. Part of the problem was that milk was no longer the local product it had been when Bromfield started dairying in the early 1940s. The nearly universal transition to bulk milk handling, combined with the nation's new interstate highway system, broke down the old geographically separate milk markets. The Borden Company stopped processing milk at its Mansfield plant in 1961, shipping milk from Malabar and other farms in the former Mansfield milkshed all the way to Columbus for processing. As the old milksheds disintegrated, dairy cooperatives consolidated, giving local farmers less control over the prices they got for their milk. The way people bought milk changed, too; by 1969, half of consumers in Columbus and Cleveland were buying milk at a grocery store instead of having it delivered to their homes. As retailers gained more power in the milk market, they further depressed the price of milk, often using it as a "loss leader"—selling it below cost to draw customers into the store and making up the loss on other products.[12]

It didn't help that milk itself was being viewed with increasing suspicion during the 1960s, as Americans became increasingly aware of the dangers of chemical contamination. In the early 1950s, concerned mothers feared that the milk they were giving their children might be contaminated with antibiotics used to treat mastitis or with fat-soluble pesticides like DDT. A complete ban on DDT use on dairy cattle in 1953 and a requirement that farmers not sell the milk from cows recently treated with antibiotics mostly addressed these issues, though some consumers worried that not all farmers complied with the regulations. Then another threat emerged—milk was often contaminated with radioactive elements like strontium-90 or iodine-131, blown onto midwestern farm fields from nuclear testing out in Nevada. Mothers feared that giving their children milk might increase their risk of leukemia, and milk sales plummeted. The immediate danger of fallout subsided when the United States and Russia signed the Partial Test Ban Treaty in 1963 and ended atmospheric nuclear testing, but that didn't help dairy farmers. Milk was no longer regarded as a safe and healthy food, and per-capita milk consumption continued to decline.[13]

In 1962, the Louis Bromfield Malabar Farm Foundation "came reluctantly to the conclusion" that it just couldn't raise enough money to fund the educational and research program it had originally envisioned. By September, the foundation could no longer afford to pay Floyd Chapman's salary, and he was asked to resign as a cost-cutting measure. Many people were sad to hear that the "dedicated, likeable, ambitious" Chapman would no longer be working at Malabar; he had done an amazing job with the limited resources available. After leaving Malabar, Chapman worked for George A. Miller at his greenhouse and floral shop in Mansfield. A couple years later, he left Mansfield to work as a horticultural therapist at the Harding Hospital in Worthington, near Columbus, where he designed and built a nature trail along a stream and helped fight water pollution in the area. After retiring from Harding in 1978, Chapman volunteered to help plan and plant Inniswood Gardens near Westerville.[14]

Chapman's resignation made many people wonder how much longer Malabar was going to survive. The *Mansfield News-Journal* pointed out that one reason people didn't donate money to Malabar was that they weren't sure exactly what it would be used for. "Is an ecological center the real goal at Malabar? Or a tourist mecca? Or both? Or something entirely different?" the editor asked. It was a valid question, but the foundation didn't answer it in the lengthy statement it released to counter what it perceived as negative publicity. Instead, it argued that Malabar was "not in any financial difficulty" and that "its current assets to current liabilities ratio is 8 to 1." Leaders did, however, admit that the foundation was "unable to carry out fully the kind of a program originally contemplated, or even all of that part of it which had already been developed" until more funds appeared. In the meantime, they encouraged people to purchase memberships in the foundation and go on tours of the farm and Big House.[15]

Without Chapman's excellent leadership, Malabar Farm entered a period of decline for the remainder of the 1960s. It took the board of trustees several months to find a new director—A. W. Short (1899–1975), who started work in May 1963. Short (who always went by "A.W.," never his full name, Alexander Walker) had been involved in the Ohio conservation movement for a long time and had served with Bromfield on the Ohio Agriculture-Conservation Commission. Short earned a BS in agriculture from West Virginia University and an MS in agriculture from the Ohio State University in 1927. He taught vocational agriculture at Hillsboro High School until 1937, when he joined the Ohio Department of Conservation. After twenty-three years of service there, he became assistant director of the Ohio Department of Agriculture. He had just retired in 1963 when he took the post at Malabar. "We are fortunate to obtain the services of such a highly-qualified, experienced and enthusiastic educator and administrator," said Ralph Cobey.[16]

Soon after Short was hired, the foundation made another major announcement: it had finished remodeling the old Schrack house into a restaurant, the Malabar Farm Inn. The cooking, featuring "just plain good home-cooked food," was done by Agnes Schwartz (1911–2008), who had previously operated a restaurant called "Joanie's Lunch" in Lucas. Helping her "on the sidelines" was E. B. "Butter" Howard, who had been appointed as a trustee of the Louis Bromfield Malabar Farm Foundation after Bill Locke passed away in November 1961. Howard, nicknamed "Butter" since childhood because his father owned a creamery, was well-known in northern and central Ohio for his cooking. He and his wife restored the historic Headley Inn on their 800-acre farm in the 1930s and operated it as a restaurant until 1962, serving traditional American dishes like corned beef, baked corn, smoked turkey, cornbread sticks, and date pudding. Headley Inn hams and bacon, cured and smoked with traditional recipes, were shipped across the country. Howard often catered barbecues for meetings, and reporters looked forward all year to the salads he served at the annual press luncheon before the Ohio State Fair. He was also active in many civic organizations, including Friends of the Land, and had been a good friend of Louis Bromfield.[17]

After the foundation put in a parking lot and received approval from the Regional Planning Commission to rezone the property for commercial use, the Malabar Inn restaurant was opened just in time for the twenty-third annual soil and health conference, held at Malabar on July 13–14, 1963. Two hundred people came to hear keynote speaker Fairfield Osborn and others discuss the dangers that pesticides posed to wildlife and human health. The emphasis on pesticides was not coincidental; this was the first institute held after Rachel Carson published her extensively researched, well-written, and disturbing critique of indiscriminate pesticide use—*Silent Spring.* Rachel Carson was very familiar with the struggles the conservation movement was facing. She had worked in a federal conservation agency (first the Bureau of Fisheries and later the Fish and Wildlife Service) throughout the 1940s and watched with concern in the 1950s as formerly influential conservationists died, retired, or were marginalized. "It is not my contention that chemical insecticides must never be used," Carson argued in *Silent Spring*. "I do contend that we have put poisonous and biologically potent chemicals indiscriminately into the hands of persons largely or wholly ignorant of their potentials for harm." *Silent Spring* reopened the public debate on pesticide safety and simultaneously revived the struggling conservation movement after a decade of intentional neglect.[18]

While many of the ideas that Malabar Farm stood for began to regain popularity in the 1960s, the farm itself continued to struggle with the same financial problems. Plans for the ecological center were permanently shelved, even as the word *ecology* began to filter into the American vernacular. Unlike Floyd Chapman, A. W. Short, an older

Sign welcoming visitors to Malabar during the Louis Bromfield Malabar Farm Foundation years. It reads: "Visitors welcome. Louis Bromfield Malabar Farm Foundation, Lucas, Ohio. Dedicated to: Better farming through conservation, education & research. Big House open daily. Sunday wagon tour. Nature trail. Ecological library. Souvenir shop. Produce market." (box 88, folder 1531, Louis Bromfield Collection, Ohio State University, courtesy of Ohio Department of Natural Resources)

man enjoying his retirement, did not keep up with the current status of ecology and the conservation movement. He preferred to focus on "spiritual and intellectual" values at Malabar. "For many people, we have learned, one of the chief things they enjoy at Malabar is just to sit on the front lawn and let their souls catch up with their bodies," he told a reporter in July 1963. When Short took over the monthly *Malabar Farm Newsletter,* the amount of actual news reported declined precipitously. Instead, he devoted most of the newsletter space to expounding his "Malabar philosophy," which was that the Malabar Farm Foundation would succeed simply because it was built on the "unselfish motives" of promoting the wise use of natural resources. The Malabar philosophy, Short believed, could "help withstand communism, hunger, poverty, and promote the general welfare and economy of all people." "We are pointing the way to a better world for people to live in," he proclaimed. "So Malabar Boosters, let's just keep on, keeping on." And "keeping on" was about all the Louis Bromfield Malabar Farm Foundation managed to do for nearly a decade.[19]

During the 1960s, Malabar stayed open as a tourist destination, museum, and modern working dairy farm. As soon as I-71 was opened in 1960, the foundation put up "prominent signs" directing travelers from the freeway to the farm. Visitors from the local area and all over the world toured the Big House, and the restaurant in the Stagecoach Inn was "noted for its outstanding 'country style' menu and succulent food preparation." The foundation sponsored two ox roasts for media representatives in the fall of 1963, featuring wagon tours of the farm and a meal prepared by "Butter" Howard. Most of the reporters who attended were impressed, and articles about Malabar's history appeared in numerous Ohio newspapers shortly after the events. The foundation continued to host annual soil and health conferences, but Short's opinion of the conferences was that the "same handful of people" attended every year and didn't care much about Malabar Farm. He questioned whether the conferences were worth "all the cost and planning," especially after only a hundred people showed up in 1964 and he sold "a lousy" two memberships and "$28.00 worth of tickets for Wagon Tours during the 2 days of the Convention." The 1965 soil and health conference attracted only seventy-five people, and newspapers didn't report on any soil and health conferences after 1968.[20]

In July 1964, the foundation was shocked to hear that beloved friend "Butter" Howard had died suddenly of a heart attack in Seattle, just after attending the Republican National Convention in San Francisco. That October, A. W. Short also had a heart attack and was taken by ambulance to Riverside Hospital in Columbus. Ralph Cobey had a heart attack in the fall of 1964 and was in the hospital again in December

Malabar Inn restaurant, 1960s. (Malabar Farm archives, courtesy of Ohio Department of Natural Resources)

1966. Jonathan Forman had a heart attack in 1965. Short's heart condition was quite serious; on February 16, 1966, his doctor told him that he only had three months to live. He opted to go through open heart surgery at the Cleveland Clinic and was in the hospital for seventy-three days. The operation, though successful, seriously damaged his voice; when he returned to Malabar in May and tried to lead tours again, he couldn't talk over a whisper. With all of the foundation's leaders suffering from such serious health problems, it's not surprising that they dropped publication of the *Malabar Farm Newsletter* in 1965. Members didn't receive a newsletter or any other regular mailings from the foundation until Harold Friar, a biology teacher from Galion who became assistant director to help out Short, resumed the newsletter in 1969.[21]

The one bright spot in 1965 was that the Malabar Inn restaurant, which had closed after the death of "Butter" Howard, reopened under new management. The building was leased to Polly Kunkle (1911–2004), who had formerly operated the Atmosphere Restaurant in Mansfield. Kunkle, "happy, moody, joyous and creative in fits and starts," soon became part of the Malabar Farm experience. She served "family-style dinners" featuring dishes like Amish baked chicken, baked steak, and "togetherness salad." All of the bread, noodles, pastries, biscuits, and corn fritters served in the restaurant were made from scratch. There was no menu for lunch; Kunkle made whatever she wanted to, and the guests usually liked it. "Her Sunday-cabbage converts are legion," one reporter noted. Polly Kunkle was most famous for her grasshopper pie, made from a carefully guarded secret recipe. All she would reveal about the ingredients was that it contained mint. News of Polly Kunkle's excellent cooking soon spread, and a large percentage of visitors to Malabar stopped at the restaurant during their trip.[22]

Despite the Louis Bromfield Malabar Farm Foundation's struggle to survive the 1960s, the number of visitors to the farm increased every year, from 15,000 in 1965 to 20,000 in 1969. Travel guides by companies such as Columbia Gas of Ohio and Nationwide Insurance listed Malabar as a tourist attraction, and the Ohio Department of Natural Resources's *Wonderful World of Ohio* magazine featured it in 1966. By 1967, so many people were coming to Malabar that the tour guides were getting overwhelmed. The first tour of the day started at 9:00 A.M., the last one ended at 5:15 P.M., and it was usually 6:00 P.M. before they could close the gift shop. Five bus tours plus another two hundred people came on one summer Sunday in 1967, and almost everyone who went on a house tour ate a meal at the Malabar Inn. Visitors saw more on their tours of the Big House, too; the girls' bedrooms and kitchen were opened to the public in the late sixties. In 1969, the gift shop was moved to a new, larger location in the remodeled Big House garage. There was so much demand for Bromfield's out-of-print farm books in the gift shop that the foundation had trouble finding copies to sell. "All evidence seems to indicate that there is a renewal of interest in Louis Bromfield," Harold Friar observed in 1969.[23]

One reason so many people visited Malabar was that finally, fifteen years after Bromfield and Friends of the Land had seen it as the wave of the future, *ecology* was becoming part of the American vernacular. In an explosion of concern about human impacts on the environment, the "ecology of man" advocated by Paul Sears fused with the new discipline of ecosystem ecology. In Cleveland, the Cuyahoga River was so polluted that it actually caught on fire, and many people predicted the imminent "death" of Lake Erie from excessive pollution. A new wave of books such as *Our Synthetic Environment* by Murray Bookchin, *Science and Survival* and *The Closing Circle* by Barry Commoner, and *The Population Bomb* by Paul Ehrlich ushered in a new era in American history—the environmental movement. On January 22, 1970, the National Environmental Protection Act, passed by both houses of Congress with only twenty dissenting votes, was signed into law. The Environmental Protection Agency was established in December 1970 with overwhelming bipartisan support. On April 22, 1970, over 20 million people celebrated the United States' first Earth Day, with speeches, protests, demonstrations, and rallies at more than 1500 colleges and 10,000 schools across the country. Although *environmentalism* was not exactly synonymous with *conservation* and some older conservationists had mixed emotions about the new movement, there is no doubt that the environmental movement helped revive interest in Malabar Farm.[24]

In the first heat of excitement surrounding Earth Day, many younger people were surprised to discover that Louis Bromfield had "preached ecology at a time when most Americans hadn't even heard the word." Dusting off Bromfield's books, they discovered that he had been concerned about environmental issues like pollution and indiscriminate pesticide use and that "in 1970 his warnings from 1940 have a prophetic ring." A revitalized and growing organic farming movement also discovered and read Bromfield's books in the 1960s and '70s, especially after Ballantine Books reissued *Pleasant Valley* and *Malabar Farm* in paperback around 1970. Many organic farmers visited Malabar Farm, often stopping at other organic farming landmarks in the same trip—like the Rodale Institute in Emmaus, Pennsylvania, and Walnut Acres in Penns Creek, Pennsylvania, the original mail-order organic food company, founded by Paul and Betty Keene in 1946. Even if they were somewhat disappointed that Malabar Farm wasn't being run exactly the way it had been in Bromfield's day, they still enjoyed touring the Big House and seeing the fields where the things Bromfield wrote about had actually happened. Both environmentalists and organic farmers downplayed any inconsistencies between Bromfield's opinions and their own beliefs, assuming that "if he were alive today" Bromfield would wholeheartedly embrace the environmental movement's emphasis on regulation and organic farming's complete avoidance of chemical fertilizers.[25]

Despite the fact that it was bringing them increased business, the Louis Bromfield Malabar Farm Foundation was slow to emphasize Bromfield's connection to the environmental movement. Friar started writing about ecology in the *Malabar Farm Newsletter* in June 1970, a couple months after Earth Day. "We are certain that Louis Bromfield would be in the thick of things," he wrote after looking at some of the books in the ecology library. But it wasn't until Ralph Cobey was appointed to the Governor's Task Force on Environmental Protection in March 1971, almost a year after Earth Day, that the foundation renamed Malabar the "Bromfield Ecology and Environmental Center." Malabar, Cobey pointed out, had great potential as an educational center, with "the largest ecology library in the world." What he didn't mention was that, in spite of the increased visitation, Malabar Farm was in worse financial shape than ever, with a net loss of nearly $15,000 from the combined farming and tourism operation in 1969. The only way Malabar Farm was even staying open was by putting off paying bills, with an occasional bailout from Ralph

Cobey's personal finances. Obviously, the Louis Bromfield Malabar Farm Foundation couldn't continue forever with this kind of deficit—especially since some people at the Noble Foundation were starting to wonder why Malabar hadn't paid a penny on the overdue loan.[26]

Even though the promissory notes for the Malabar loan had been due in 1966, the Noble Foundation had not really expected the Louis Bromfield Malabar Farm Foundation to pay it back that early. But the two parties did not renegotiate the loan in 1966 as planned; instead, the overdue mortgage continued to accrue 4 percent interest, compounded annually. A few years later, the Noble Foundation went through some major reorganization to comply with the 1969 Tax Reform Act. The end result, achieved in 1972, was that a new publicly held corporation called Noble Affiliates, Inc. was created to fund the Noble Foundation, allowing it to become a granting entity in addition to a research institution. At the same time, however, the new law meant that the Noble Foundation could not have an outstanding overdue loan to Malabar Farm; Noble either had to foreclose or renegotiate. In 1970, Ralph Cobey started looking for someone to take over the farm and pay off the mortgage. He started conversations with the Association of Ohio Commodores, the Technical Institute District in Ashland-Crawford-Richland Counties, the Battelle Institute, and even the Borden Company. But all of these negotiations fell through, and in September 1971 the Noble Foundation told Cobey that it would be forced to foreclose "unless other satisfactory arrangements are made immediately." "The IRS will not accept conversations with the debtor in lieu of positive effective action to collect the debts due," they explained.[27]

Probably in an attempt to get Cobey to do something more than have conversations, John March, Noble Foundation president, alerted the local newspapers to the possibility that Malabar might be foreclosed. "There is no malice in this action," March explained. "It is just a matter of business." He explained that the Noble Foundation was "in sympathy with what they are trying to do at the farm," which was why it had "gone along for six years without forcing any payments on the note." But, if nothing changed, Noble would be forced to foreclose "by the restrictions in the 1969 Tax Reform bill." Not surprisingly, Ralph Cobey was pretty upset when he saw the story in the newspaper. "I just don't understand, this was so important, why I wasn't told about it," he told Sam Noble. "After all, Sam, to begin with I was just as innocent in this Malabar thing as you are." Sam replied that it would "certainly be desirable" if the Louis Bromfield Malabar Farm Foundation sold part of the land to pay the Noble Foundation back, or sold the farm to "some responsible group." "However, if all of the above fails, the Foundation has no alternative but to foreclose," he concluded.[28]

Once again, the future of Malabar Farm was in jeopardy. If Ralph Cobey would have had his way, no one would have known about it until it was too late. But once the plight of Malabar became public knowledge, it quickly became evident that far more people cared about Malabar than had in 1958—and they were willing to fight to save the famous farm.

*Chapter 10*

# For the People of Ohio

"The possibility that Malabar Farm would not continue to be operated as an outstanding example of conservation, as a facility available for the public to see and experience, and as a memorial to Mr. Bromfield, is extremely distressing," Ohio Department of Natural Resources (ODNR) director William B. Nye wrote to the Noble Foundation in September 1971. Like most other people in Ohio, Nye was shocked when he heard that Malabar might be foreclosed. He explained that ODNR was interested in managing Malabar Farm as a joint operation with the Ohio Department of Agriculture but that the agency had no budget to purchase the property. D. K. Woodman, editor of the *Mansfield News Journal,* wrote to the Noble Foundation explaining that he understood why it had to foreclose on the defaulted mortgage but hoped that they would either operate the farm themselves or give it to the State of Ohio. Meanwhile, Carmen Stricklen, who worked as the Louis Bromfield Malabar Farm Foundation's secretary, reported that real estate agents were "all over the place" at Malabar Farm. "If Malabar closes, I'll be just sick," she said. The fate of Louis Bromfield's world-famous farm was in the hands of the Samuel Roberts Noble Foundation, in far-off Ardmore, Oklahoma—and no one knew whether it would donate the farm "for the good of the people of Ohio" or sell it to developers after it foreclosed.[1]

The first time Malabar had been threatened with development, in 1957, only aging conservationists had fought to save the farm. But this time, at the height of the environmental movement, a new generation of young people rallied to save Malabar. "Malabar Farm, the unique 600 acre example of natural ecology, could die August 1 and be replaced by real estate developments," warned a group of students from Kent State University. "The untouched beauty of this farmland *must* not become a concrete graveyard." The Kent State students believed the best way to save Malabar was to turn it into a national park before it was "crushed beneath the bulldozers and bricklayers of a blind society." They formed a Committee to Save Malabar in April 1972, sent letters to their senators and representatives, and circulated petitions to put the farm into "national trust."[2]

Lila Wagner, a fifth grade teacher at Ledgeview Elementary School in Macedonia, attended some of the Kent State meetings and realized that the group's proposal to turn Malabar into a national park was not going to work. After some thought, she came up with a different plan. School children in Massachusetts, she recalled, had successfully raised funds to preserve the historic battleship Old Ironsides. Why not try the same thing with Malabar? Along with Ruth Colton, a retired teacher in the area, Wagner launched the Children's Crusade to Save Malabar. The students at Ledgeview put on a "Malabar Bazaar" and bake sale, made hundreds of Malabar yo-yos to sell, and spent hours addressing letters and putting them in envelopes to mail to every other elementary school in the state. Children across Ohio donated their pennies and nickels to save Malabar, netting about $6,500.[3]

While neither the Kent State campaign nor the Children's Crusade raised enough money to pay off the $200,000 overdue mortgage and interest, they both were very effective in raising public awareness about Malabar. "We don't want to be painted as a bunch of villains out here," explained John March, the president of the Noble Foundation. "We saved Malabar from going under to real estate developers once before in 1956. We're trying to do that again." In June, the Noble Foundation agreed to forgive the mortgage and donate the farm to the State of Ohio. "Through letters, countless newspaper and magazine articles and through the donations of private funds, thousands of Ohioans have indicated the desire that this conservation and historical gem be preserved for the enjoyment and education of future generations," announced Governor John Gilligan on June 14. "With the property coming under ownership of the citizens of Ohio through their government, the preservation can be assured." After some negotiations, including an agreement that the State of Ohio would pay Ralph Cobey $26,436.94 for his share of the mortgage, the Louis Bromfield Malabar Farm Foundation deeded Malabar Farm over to the Noble Foundation to satisfy federal tax laws. Then the Noble Foundation, as a granting institution, donated the farm to the State of Ohio to use "as a site for ecological research and the development of sound conservation practices in agriculture." For the third time, Malabar was saved by the Noble Foundation—this time permanently.[4]

The deed to Malabar was officially transferred to the State of Ohio on August 3, 1972. Over two hundred people, including Ellen Bromfield

Malabar Farm as it looked in 1973, shortly after the state took over. (Malabar Farm archives, courtesy of Ohio Department of Natural Resources)

Geld, attended the formal ceremony on the lawn in front of the Big House. "We understand the great responsibilities that go with this gift," Governor Gilligan said as he accepted the deed. "We will do our very best to preserve what has been so beautifully preserved up to now. We promise to preserve Malabar Farm for generations to come." The state agreed to have the Louis Bromfield Malabar Farm Foundation continue operating the farm until October 1, at which time the Department of Natural Resources and the Department of Agriculture began jointly managing it. Along with the Malabar property and Big House furnishings, the State of Ohio also took over the popular Malabar Inn restaurant, which was still being operated by Polly Kunkle. When she first heard that Malabar was being transferred to the state, Kunkle worried that "a quick-serve hamburger stand might be brought in to replace her." "Good heavens, no!" exclaimed Gene Abercrombie, director of the Ohio Department of Agriculture. "Why, everything at Malabar is legend. So is Polly Kunkle; she's our newest." Even before the official deed transfer took place, her lease on the Malabar Inn was extended for another three years.[5]

Shortly after the state received title to Malabar Farm, Governor Gilligan formed the Governor's Advisory Council on Malabar Farm to help plan how to best manage the land. The advisory council included many people who cared about Malabar, all the way from old friends of Bromfield like Jonathan Forman and Max Drake to campaigners to save the farm like Lila Wagner and Carmen Stricklen. While everyone on the council wanted to preserve and use Malabar for the good of the people of Ohio, they had different ideas about how to reach that goal. Max Drake wanted the farm to be a showcase of modern farming, with the animals moved away from the Big House to cut down on smells and congestion. Jack Miller, chief of ODNR's Division of Parks and Recreation, thought Malabar would be a good model farm to expose inner-city kids "to a farm-type environment" and show them where their food came from. Several people, including Ralph Cobey, suggested constructing a museum "and other educational facilities at Malabar." Gene Abercrombie suggested setting up a petting zoo and letting the USDA use five acres of the farm as a "test plot" for "trees and ornamental shrubs."[6]

The State of Ohio's original plan delegated the management of Malabar's farming operations to the Ohio Department of Agriculture (ODA) and put the Big House and nonagricultural land under the jurisdiction of ODNR's Division of Parks and Recreation. To help make management decisions about the property and restore the historic structures "back to their original appearance as much as possible," ODNR hired C. J. "Bill" Solomon of Mount Vernon, one of Louis Bromfield's old friends and neighbors, to serve as the farm's curator. For the rest of the nonagricultural property, ODNR drew up several development plans. The most intensive of these involved building a ski area on Mount Jeez, damming Switzer's Run to create a twenty-acre lake and beach, building a riding stable across from the Malabar Inn, constructing a camp with cabins and dining hall "on the southern part of the farm"; and building two outdoor theaters, a farm museum, and a trout hatchery. The advisory council was opposed to this kind of intensive recreational development, which "could soon destroy the famous farm as an example of natural beauty and the best in agricultural practices" and would duplicate facilities already available at nearby parks. With so much negative feedback, most of the intensive development plans were scrapped.[7]

The council also debated whether the farming operation at Malabar should be "frozen in time" or showcase the best in modern farming methods. The final decision was up to ODA director Abercrombie, and, like every other farm manager since Bromfield's death, he decided to make the farm "prove" that modern dairy farming was profitable. It would take a lot of work to get the farm up to date, though. Due to the dire financial status of the Louis Bromfield Malabar Farm Foundation, the farming operation had become pretty rundown by the time the state took over. The foundations of the barns were crumbling, all the buildings needed paint, the dairy operation had been downgraded to Grade B, multiflora hedges were taking over the fields, and Bromfield's trench silo was filling with water and its sides were starting to cave in. To get the farm back in shape, ODA hired a young man named Dana Bass, who had been working for ODA's Market News Service and saw the job as an opportunity "to show myself and others what I can do." His goal was to get the dairy operation back to Grade A status and make it "the top herd in Richland County."[8]

One of the biggest problems that the state had to deal with was the out-of-control multiflora rose hedges, which Abercrombie called "Bromfield's biggest mistake." Since Bromfield's death, multiflora rose had spread into the farm's pastures and fields and the Big House's backyard. While the hedges delineating the fields were left in place for the time being because Bromfield had planted them, an aggressive campaign to clear the invasive plant out of the rest of the farm was launched. As workers hacked away the thorny tangle behind the Big House, they discovered a "fireplace, swings and tennis courts" originally constructed by Bromfield "for his family and personal friends." A more controversial part of the farm modernization program was Abercrombie's decision to demolish Bromfield's trench silo, originally constructed in 1946. On July 9, 1973, a bulldozer arrived on the farm and destroyed both the trench silo and a small tower silo near the main dairy barn. Abercrombie assured newspaper reporters that removing the trench silo "won't change the appearance of anything," but Bill Solomon disagreed. "It was as much a part of Malabar as the Malabar name itself," he mourned. To

By 1973, the multiflora rose hedges at Malabar were out of control. (Malabar Farm archives, courtesy of Ohio Department of Natural Resources)

replace the trench silo, ODA spent $10,000 to build a large concrete stave tower silo next to the dairy barn.[9]

During the first two years of state ownership, things seemed to be going well at Malabar as both ODNR and ODA worked hard to clean up the grounds and buildings. Malabar was "being restored to the productivity of the prime years of Bromfield's ownership, and as a place for all Ohioans to visit," proclaimed a new state-sponsored brochure about the farm. But 1974 was a gubernatorial election year, and the farm's management became embroiled in politics for the first time ever—one of the downsides of state ownership. First, Louis Lamoreaux, the architect who designed the Big House, claimed that many items that had been at Malabar during Bromfield's lifetime were "missing." ODNR investigated Lamoreaux's claims and discovered that he had based them on photographs from the early 1940s; photographs taken shortly before Bromfield's death showed that nothing had disappeared during state ownership. Less easy to resolve was a controversy about a half-acre parcel of land adjoining Malabar that Bill Solomon bought at the advice of ODNR director William Nye and planned to resell to the state for a few thousand dollars more than he paid for it. State representative Joan Douglass, a realtor, claimed Solomon was making a profit on the deal; Solomon insisted that he was only covering his expenses for holding the property for a year. That tiny parcel of land became so controversial that the state never did buy it from Solomon, even when he sued them a few years later.[10]

In early 1975, when the new administration took over, a much more serious allegation was raised against the state's management of Malabar. An investigation of "financial irregularities" in the farming operation discovered that the farm manager, Dana Bass, had sold cattle, bull semen, and grain from the farm and pocketed the proceeds. Bass resigned at the beginning of the year and moved to Mansfield. When the Richland County Grand Jury indicted him for theft, Bass pleaded innocent but gave a $2,700 check to John Stackhouse, the state agriculture director. When the probe into Malabar's finances was complete, investigators discovered that Bass had stolen a total of $10,071. Faced with the evidence, Bass finally pleaded guilty on August 22, 1975.[11]

The new ODNR director, Robert Teater, called the operation of Malabar Farm a "mess." He blamed most of the farm's problems on the conflicting management plans of ODA and ODNR and suggested that "one department should run the whole operation." On February 5, 1976, Governor James A. Rhodes announced that sole control over Malabar would be given to ODNR. The farm would be "operated primarily as an historical, recreational and educational center," and the farming operations would "not be emphasized." As part of the plan to turn Malabar into a full-fledged state park, ODNR decided to purchase a couple hundred adjacent acres of land to serve as a buffer zone and to increase the park's recreational value. Some property owners were glad to sell, but others protested. The land acquisition soon got tied up in politics, with the Democratic State Controlling Board refusing to release funds for land purchase until an audit was run on the Republican ODNR's operation of Malabar Farm. The controlling board finally approved the land purchase in late November, allowing ODNR to add 198 more acres to the property. The audit, which looked at the state's management from 1972 to 1976, failed to find any evidence of mismanagement but did confirm that the farming operation had sustained a net loss of $143,434 over four years, despite all of ODA's plans for making the dairy pay for itself.[12]

To get Malabar back on track and make it a park worthy of Louis Bromfield's legacy, ODNR appointed a twenty-six-year-old park naturalist named James M. Berry to be the new park manager. Berry graduated from the School of Natural Resources in OSU's agricultural college in 1972 and had worked as a naturalist at Mohican, Malabar, and Hueston Woods state parks. At Hueston Woods, Berry had helped plan events like a maple syrup festival and a popular living history program called "Heritage Days," where costumed reenactors demonstrated historic skills like dipping candles, making quill pens, carving wood, and dying fabric using natural materials. These historical programs reflected a huge resurgence of interest in pioneer history all across the United States, stimulated by the nation's approaching Bicentennial. Established historical sites

Jim Berry at Malabar, 1976. (Courtesy of Jim Berry)

like Plimoth Plantation, Colonial Williamsburg, and Greenfield Village changed their interpretive programs to include more live cooking, crafting, and agricultural demonstrations, while new living history farms were formed across the country to preserve vanishing remnants of America's agricultural heritage.[13]

As the only historic farm in the Ohio state park system, Malabar took on new significance in the year of the Bicentennial. Jim Berry's task was to counter negative publicity with a positive and relevant set of programs. The entire department, including ODNR director Robert Teater, Division of Parks chief Don Olsen, and deputy chief Larry Henry, gave Jim Berry the encouragement and support he needed to turn Malabar Farm into one of the most program-intensive parks in the entire state. "Programming is what will prove Malabar's worth," Berry said in 1978. "Farming has always been here and we have improved on it, but the programs are what serve the public." With ODNR's emphasis on programming, "Malabar took off like a shot." In just three years after ODNR took over sole management of Malabar, park attendance quadrupled from 1976 levels.[14]

In celebration of the Bicentennial, Jim Berry's first major assignment as park manager at Malabar was to plan a statewide Ohio Heritage Days festival. On a summer afternoon, he sat down on the chairs on the Malabar lawn with two of ODNR's regional naturalists, Howard Gratz and Chris Arn, to plan the event. Held on November 6–7, 1976, this first Ohio Heritage Days celebration was publicized as "a taste of life in the 'good 'ole days' on America's 'first frontier'" and a "trip back in time." Costumed volunteers demonstrated historic skills like spinning, weaving, dyeing, candle-making, soap-making, shoe-making, quilting, hide-tanning, and corn-shelling. Visitors sampled freshly churned butter, smoked meats, maple sugar candy, and "pioneer teas and jellies." They looked at displays of antique farm tools and machinery and covered their ears while reenactors shot off "old black powder guns." Park staff offered hayrides, wagon tours, Big House tours, and an evening barn dance. Despite the cold November weather, over five thousand people attended, and the event was a huge success.[15]

Ohio Heritage Days quickly became the biggest annual event at Malabar Farm. Starting in 1977, ODNR moved the event to the third weekend of September so that it would be more enjoyable to be outside. Over fifteen thousand people flocked to the farm that year to watch costumed craftspeople demonstrate making forty different kinds of items, ranging from sauerkraut to cedar-shake shingles. They sampled corncob jelly, apple butter, and sassafras tea cooked in huge kettles over open fires. Kids could carve their own apple-head dolls and see the Borden cow Elsie and her calf Beauregard, while adventurous adults tried smoking

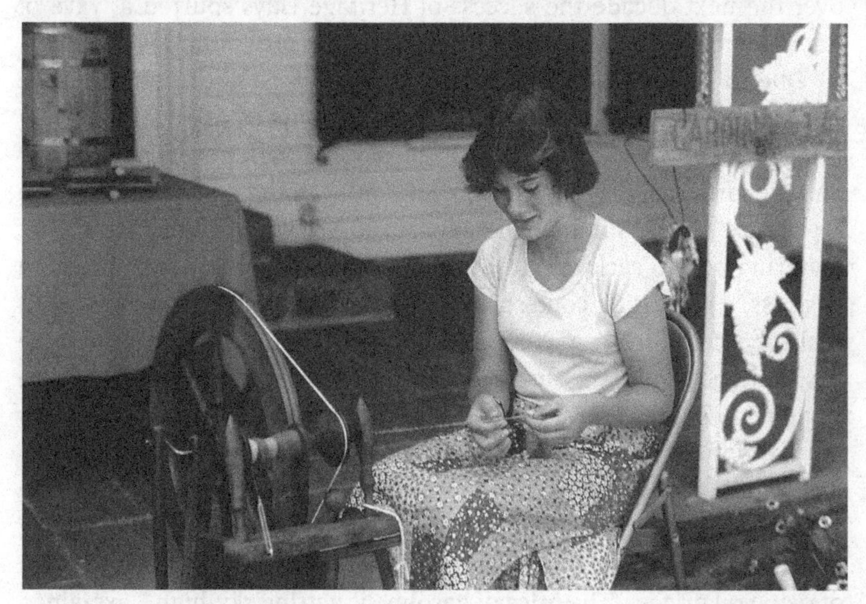

*Above:* Parked cars at Heritage Days, 1981. (Courtesy of Jim Berry)

*Left:* Spinning demonstration at Heritage Days, 1978. (Courtesy of Jim Berry)

"wild tobaccos" like sumac and corn silk. Sixty thousand people attended in 1980, and seventy-five thousand came in 1982 to see and participate in over a hundred demonstrations and activities emphasizing "rural pioneer skills." "It was huge," Jim Berry recalled. "Rangers came from state parks all over the state to handle the traffic, parking, and crowds. It was all free. It was a logistical nightmare, but we were able to pull it off. It grew like crazy." Ohio Heritage Days was one of the biggest

Draft horses at Malabar, 1980s. (Courtesy of Jim Berry)

festivals sponsored by the State of Ohio, second only to the Ohio State Fair. More than any other single event, it revitalized Malabar Farm, and over the next decade the success of Heritage Days spurred a wave of improvements in Malabar's infrastructure and programming.[16]

Due to the intense popular interest in pioneer history in the wake of the Bicentennial, many of the programs launched at Malabar during the late 1970s were historically oriented. The park put a collection of antique farm tools and machinery on display in the dairy barn near the Big House. In the summers, Malabar held an old-fashioned Farm Day, featuring potato sack and three-legged races, a greased pig chase, an egg toss, a pie-eating contest, and even a tobacco spitting contest with real tobacco. Jim Berry partnered with the Central Ohio Draft Horse Association to host the first-ever Spring Plowing Days in 1979, where farmers used draft horses and plows to prepare several acres at Malabar for planting. Some of the three thousand people who attended the second annual Spring Plowing Days in 1980 were merely interested in the event from a historical perspective, but others were seriously considering draft horses as an economically viable alternative to tractors in an era of rising oil prices. "The price of gasoline is getting sky-high," explained Bill Reed, the president of the Central Ohio Draft Horse Association.[17]

The interest in pioneer farming methods in the 1970s was more than mere nostalgia. While most people in Louis Bromfield's time had dismissed pioneer agriculture as wasteful and inefficient, the energy crises of the 1970s highlighted the inherent vulnerability of modern agriculture's dependence on fossil fuels. A combination of political and market factors, Middle Eastern unrest, and the inability of domestic oil production to keep up with consumption caused oil crises in 1973 and 1979. Panicking drivers waited in lines for hours at gas stations to get less gas at higher prices than they were accustomed to, and a related heating oil shortage meant that some homeowners could not get enough fuel to heat their homes during the winter. Faced with imminent shortages of oil, natural gas, and even electricity, many people took seriously the computer model predictions of the 1972 *Limits to Growth* study, which projected that industrialized civilization would "overshoot and collapse" because of resource depletion or pollution "well before the year 2100." Over a million young people in the 1970s, inspired by books like Helen and Scott Nearing's *Living the Good Life,* felt their best hope of surviving this crisis was to go back to the land and establish self-sufficient homesteads. Historical skills that used hand tools and natural materials weren't just relics; they were potentially life-saving alternatives to increasingly unstable and polluting modern technologies.[18]

Along with a renewal of interest in historic farming methods, the oil crisis stimulated an intense interest in all forms of alternative energy. One popular suggestion for lessening the United States' dependence on imported oil was to run the American fleet of automobiles on "gasohol"—a mixture of 10 percent corn ethanol and 90 percent gasoline. With so much public interest in ethanol, Jim Berry got a permit to build a still at Malabar and run one of the park's tractors, a 1950 International M, on gasohol. He tried building a solar-powered still from plans published in *Mother Earth News,* but the design didn't work, so he built a wood-burning still instead. After running the tractor on gasohol for a few months, the park staff discovered that the homemade ethanol was starting to corrode the fuel lines and switched back to normal gasoline to prevent further damage.[19]

Malabar ethanol still, 1979. (Courtesy of Jim Berry)

The most successful alternative energy event at Malabar was Wood Heating Days, which was launched in 1978 and continued for about five years. Wood-burning heating stoves appealed to people in the late 1970s for several reasons—historical interest fanned by the Bicentennial; shortages of heating oil, natural gas, and sometimes even electricity in the energy crisis; and the attraction of using the "effective, cost-saving, renewable resource" of wood for self-sufficient heat. Berry combined all three of these interests in an all-day annual event. The first Wood Heat Day, in 1978, featured wood-splitting demonstrations, a display of various models of wood stoves, and "an old-fashioned bean lunch, cooked over an open fire." By 1980, Wood Heating Days had grown to a weekend-long event in early November, featuring vendor displays of twenty-five wood stove models, log skidding demonstrations by the Central Ohio Draft Horse Association, a lumberjack show featuring log-rolling and sawing contests, and chain saws and log splitters in action.[20]

In response to widespread public interest in homesteading, Malabar Farm State Park offered a series of self-sufficiency workshops between 1981 and 1984. At these monthly events, participants could learn how to spin yarn from wool, manage their woodlots sustainably, grow organic gardens, can and dry fruits and vegetables, keep bees in their backyard, forage for wild foods, and make their own Christmas decorations from natural materials. Most of the self-sufficiency classes were held in the Louis Bromfield Learning Center, a log cabin built by Bromfield's friends and neighbors Jim and Georgia Pugh in 1938 and purchased by ODNR in 1977. The cabin, which Jim Pugh had constructed from retired utility poles that he had shaved down with a drawknife, was decorated with bronze chandeliers and stair rails salvaged from old buildings in Mansfield and featured a huge sandstone fireplace, perfect for the popular hearthside cooking classes. Even though it was only a couple years older than the Big House, the Pugh cabin fit perfectly with the park's historic theme. Along with the upper Pugh cabin, the state also purchased a second cabin just down the hill, which Pugh built in 1956.[21]

Just across the old Ferguson Road from the upper Pugh Cabin, the park constructed a sugar shack for the second most popular event of the year—the Maple Syrup Festival. This festival started in 1978, when Jim Berry bought a small hobby evaporator and tapped about a hundred maple trees. Park staff and volunteers educated visitors about the history of maple syrup making by demonstrating various methods—from filling hollow logs with sap and adding hot rocks to boil off the excess water, to tapping trees with hand-carved wooden spiles and buckets and boiling down the sap in big kettles over an open fire, to modern plastic tubing and evaporators. The festival was extremely popular, and for the 1981 season the park constructed a sugar shack, installed modern blue plastic tubing in the sugar bush, and bought a three-by-twelve-foot commercial evaporator to collect and process greater volumes of sap. The sugar shack was constructed with salvaged lumber from the state park system, milled at the new "historically authentic" Malabar sawmill, constructed in the former dairy pasture near the park boundary in 1979. The sawmill was initially powered by a steam engine for historical demonstrations and later by a diesel engine, and the lumber milled there was used for various building projects around the farm.[22]

While the historical programs held at Malabar during the 1970s spanned a much longer period of history than Louis Bromfield's lifetime, the history of Malabar itself was not neglected. Jim Berry read Bromfield's farm books and began reorienting the farm program toward "things that the public was more interested in" instead of the "ultramodern agribusiness operation" that ODA had tried to turn Malabar into. The idea of

*Left:* Sugar shack at Malabar, 1980s. (Courtesy of Jim Berry)

*Below:* Malabar farm sawmill, 1981. (Courtesy of Jim Berry)

Gallipolis Matt Fran, high-yielding Holstein cow. (Courtesy of Jim Berry)

making the Malabar dairy operation break even financially was finally abandoned. The milking herd was downsized to ten or twelve cows, but the Malabar dairy maintained its Grade A status and sold milk to a cooperative called Milk Manufacturing, Inc. As part of the enhanced educational experience, each cow in the reduced dairy herd was carefully selected for some unique characteristic. Berry diversified the all-Holstein herd to include Jersey, Guernsey, Brown Swiss, and Ayrshire cows so visitors could learn about different dairy breeds. Another attraction was Number 51 or Gallipolis Matt Fran, an "elite" high-yielding Holstein cow who gave an average of eleven gallons of milk per day—twice the production of an average Holstein at the time.[23]

Along with downsizing the dairy herd, the entire farming operation at Malabar was diversified to reflect what it might have looked like under Bromfield's original self-sufficient Plan. Chickens, ducks, turkeys, goats, a team of oxen, a Boxer dog, and two Percheron draft horses named Tom and Pete joined the Malabar menagerie. The crops grown at Malabar were selected for educational purposes, too. In 1977, Berry planted heritage and open-pollinated varieties of corn next to modern hybrids so that visitors could see "how one stalk differs from the others." He also planted rye, barley, sunflowers, and buckwheat "as educational displays for visitors" but omitted grain soybeans from the crop rotation because Bromfield had not planted them. In 1978, he planted a quarter acre of sorghum to educate visitors about this historic crop. The farm again grew vegetables, which were sold at Bromfield's vegetable stand. "Our goal has been to make Malabar a total farm experience rather than just the house tour that it once was," Berry explained. "We are still farming here, but we are doing it on a demonstration scale."[24]

Tractor-drawn wagon tours remained a popular attraction at Malabar in the 1970s and 1980s. This photo was taken on Senior Citizens' Day in 1977. (Courtesy of Jim Berry)

On the Malabar hills and fields, Berry demonstrated strip cropping, rotations, and conservation tillage. Interest in Bromfield's conservation tillage methods began to revive in the 1970s, partially because of the oil crisis and partially because a soil erosion crisis was once again sweeping the country. With all the emphasis in agriculture on economics instead of conservation, farmers in the 1960s and 1970s had ripped out their fencerows, plowed up marginal land, and neglected the tried-and-true erosion control methods developed by Hugh Bennett. New, larger tractors couldn't navigate the narrow contours originally laid out in the 1930s and 1940s, so contour plowing and strip cropping were abandoned on 90 percent of American farmland. While the big tractors and plows smoothed out incipient erosion channels before they could become gullies, the Soil Conservation Service calculated that nearly 2 billion tons of topsoil were being lost to water erosion in 1977—about two-thirds of the catastrophic soil erosion rates that had stimulated the original soil conservation movement in the 1930s.[25]

The concern about soil erosion in the 1980s took a somewhat different form than that of the 1930s, just as the sheet and rill erosion was more subtle than the old-time devouring gullies like Providence Canyon. In the environmental age, the greatest concern was the off-site impacts of soil erosion—non-point source water pollution from fertilizers, pesticides, and soil particles. The solution, too, was different. With traditional engineering practices mostly out of the picture, soil conservationists encouraged farmers to use conservation tillage—defined as any tillage system that kept at least 30 percent of the soil surface covered with crop residues. Chisel plows, descendants of the Graham-Hoeme plow used by Bromfield, made a comeback. Systems like strip tillage and ridge tillage only broke up the soil where seeds were planted. Least erosive of all was a system called "no-till," which used herbicides to completely eliminate tillage—though some people worried about the negative environmental impacts of this intensive herbicide use. Using a combination of these different conservation tillage methods, American farmers were able to decrease soil loss from water erosion by 37 percent between 1982 and 1997, down to about 1 billion tons a year. Edward Faulkner's idea of eliminating the moldboard plow had finally become mainstream, nearly forty years after he published *Plowman's Folly.*[26]

Soon after it took over sole operation of Malabar, ODNR made several infrastructure improvements to make Malabar Farm more visitor-friendly and enhance recreational opportunities. The shortage of

Aerial view showing new parking lot at Malabar (*left*), circa 1980. (Courtesy of Jim Berry)

parking near the Big House was becoming acute, and the large amount of traffic on Bromfield Road posed a hazard to pedestrians, so in 1980 the state constructed a new parking lot in the dairy pasture near the Big House, with its own entrance from Pleasant Valley Road. It also added a second parking lot and a picnic area near the working farm and put up a gate to control access to Mount Jeez. A second picnic area, horse staging area, and horse campground were constructed on the property that the state had purchased in 1976. Young men and women from Ohio's Youth Conservation Corps helped build an eight-mile horse trail and three nature trails in the Malabar woods. The "mail-order house" that Louis Bromfield had detested was remodeled and converted into a youth hostel in 1978, providing inexpensive housing for groups and other visitors to Malabar.[27]

While summer had previously been the main tourist season at Malabar, Berry worked to make it a fun and exciting place to visit all year round—even in the cold of winter. Starting in 1976, the park hosted a "Christmas at Malabar" celebration, featuring evening tours of the Big House, carolers from local churches and other organizations, freshly baked cookies, wassail, horse-drawn wagon rides around the farm, and a live nativity scene. Beginning in 1981, garden club members from Mansfield decorated the Big House with natural wreaths, pine branches, pine cones, and candles, instead of the artificial greenery previously used by ODNR. After Christmas, the park offered ice skating on the big pond behind the working farm, with hot chocolate and a campfire to warm up skaters. The park offered its first cross-country skiing workshops in 1979, advertising that it was the first state park in

Ice skating at Malabar, 1980s. (Courtesy of Jim Berry)

Ohio to have groomed cross-country ski trails. For several years, the park even offered ski rentals for people who wanted to try skiing before purchasing their own equipment.[28]

Malabar Farm State Park thrived in the 1980s as a popular destination for people interested in Louis Bromfield, farm animals, pioneer history, or just quiet hikes in the woods. Malabar Inn remained a popular part of visiting Malabar, even after Polly Kunkle retired in September 1976 because of heart issues. After extensively renovating the building, ODNR leased the restaurant to an Austrian couple named Leo and Betty Rukavina. The park sponsored events for people of all ages and interests, from the ever-popular Ohio Heritage Days to a music festival featuring dulcimers and other traditional acoustic instruments. Park ranger Reid Caldwell led efforts to save the bluebirds by putting up nesting boxes, and hundreds of people came out in April for the "Spring Volksmarsch," a ten-to-thirteen-kilometer noncompetitive hike along the park's trails. In 1981, director Michael DeLauro even filmed a twenty-eight-minute documentary about Malabar, called *A Pen and a Plow*. The film was partially funded by the money Lila Wagner and Ruth Colton had raised in their Children's Crusade to Save Malabar back in 1972, and the premiere was appropriately held in the auditorium at Ledgeview Elementary School in Macedonia. By 1985, annual attendance at Malabar was up to 180,000 people, and publicity about the farm was almost universally favorable.[29]

Programming at Malabar began to change in the late 1980s, reflecting broader changes in American society. While Heritage Days and the Maple Syrup Festival remained popular, attendance at Spring Plowing Days dropped dramatically, from three thousand people in 1980 to only 350 in 1987. Once the oil crisis was over, horse-powered agriculture no longer seemed attractive to most farmers. Global free market oil futures trading and an end to price controls, plus increased American military involvement in the Middle East, made gasoline cheap and abundant once again by the mid-1980s, despite the fact that domestic oil production continued to decline. Meanwhile, many back-to-the-land homesteaders found that there was "a gap between the idealized theory of pastoral life and the exhausting reality" and that endlessly hauling water and trying to cook on a wood stove was not as fun as they had originally thought. They also learned the hard way that it was impossible to be completely independent of the mainstream economy; they still had to earn some kind of income to pay bills. Once the economy rebounded in the 1980s and the collapse of society did not seem so imminent, many homesteaders moved back to urban areas. Those who stayed in the country often modernized their homes and gave up the elusive ideal of total self-sufficiency. By 1985, most of Malabar's self-sufficiency classes had been discontinued due to lack of interest; only the hearthside cookery workshop remained an annual tradition.[30]

While he still loved Malabar, park manager Jim Berry became frustrated with ODNR politics in the mid-1980s, especially after the Celeste gubernatorial administration took over. "It was so political that if I was to hire someone for instance, the county head of the Democratic Party would have to approve them first," he explained. "I knew I could retire with thirty years of service and lead a comfortable life, but I couldn't sacrifice my principles for that enticement." In June 1986, Berry left Malabar Farm to take a new position as director of the nonprofit Cincinnati Nature Center. His position as Malabar park manager was filled in August 1986 by Brent Charette, who had a degree in forestry from Hocking Technical College and had worked for ODNR since 1979 at Salt Fork and Paint Creek State Parks and as a statewide naturalist. "Malabar was always an amazing place for me," recalled Charette, whose favorite task as park manager was running the annual Maple Syrup Festival. Charette changed little at Malabar while he was manager, continuing the most successful programs developed by Jim Berry. The biggest event of the year continued to be Ohio Heritage Days, where steam engines from all over the state ran the sawmill, a handle-making machine, a shingle mill, and machines that harvested and threshed Malabar-grown wheat.[31]

By 1990, the number of people visiting Malabar annually was ten times higher than it had ever been during Louis Bromfield's lifetime. And finally, thirty years after his death, Bromfield's ideas about working with nature to develop a permanent agricultural system were being seriously considered once again—under the new name of "sustainable agriculture."

# Chapter 11

# Sustainable Agriculture

During the summer of 1987, a group of volunteers from the Ohio Ecological Food and Farm Association (OEFFA) revived Louis Bromfield's vegetable garden at Malabar Farm. They tilled and planted vegetables in a sixty-by-ninety-foot plot across from the Malabar Inn and demonstrated methods like raised-bed gardening, companion planting, and mulching with old hay from one of the Malabar barns. Despite a shortage of time and personnel, the "dedicated and experienced gardeners" from OEFFA still reaped a "bountiful harvest of vegetables." The volunteers made most of the vegetables into soup, which they served to over four hundred people during the annual Ohio Heritage Days festival in September. Although OEFFA could not get enough volunteers to replant the garden in 1988, their partnership with ODNR was one of the first times that Malabar was connected to a new movement that was variously known as *alternative agriculture, low input agriculture,* and, most enduringly, *sustainable agriculture.*[1]

By the 1970s and 1980s, mainstream American agriculture had become increasingly dependent on ever-larger machinery and ever-higher amounts of chemicals. The oil crisis, worries about the environmental and human health impacts of pesticides, concern about the loss of small family farms, increased awareness of soil erosion, and an ongoing farm economic crisis combined to make many people question the long-term viability of this system. In 1979, a group of concerned Ohioans formed the Ohio Ecological Food and Farm Association with the goal of building "a healthy, accountable, and permanent agriculture in Ohio." Similar groups were started in other regions, and the USDA even concluded in a 1980 study that most organic farmers were using "best management practices." Since the word *organic* still aroused emotional opposition from conventional farmers, proponents of ecological farming systems in the 1980s increasingly began using the phrase *sustainable agriculture*. Sustainable agriculture appealed to almost everyone because of its inherent ambiguity; various groups emphasized its social, environmental, soil conservation, or economic aspects.[2]

In December 1987, Congress made history by allocating a small amount of funding for research on "low input sustainable agriculture"; in 1990, the program's name was changed to "Sustainable Agriculture Research and Education." For the first time since the 1950s, agricultural scientists could research natural farming methods without being "ignored, ridiculed, and persecuted by their colleagues." The Ohio State University started a sustainable agriculture program in 1986, led by an entomologist from England named Clive Edwards, and in 1988 over 450 people attended an International Conference on Sustainable Agricultural Systems in Columbus. Despite dropping the word *organic*, the sustainable agriculture movement still faced some opposition. One antiorganic activist organization, Food for Peace, claimed that the speakers at the Columbus conference were "designing plans to set agricultural technology back hundreds of years." But few people wanted to be against sustainability, and from the 1990s on, even the chemical and agribusiness industries worked hard to portray themselves as "sustainable."[3]

As the sustainable agriculture movement grew, many people rediscovered its ideological connection to Louis Bromfield and Malabar Farm. Organic farmers had been reading Bromfield's books for years, and most of the Ohio farmers who formed OEFFA "knew about Bromfield and Malabar." But Bromfield's farming methods, especially his use of chemical fertilizers, had never fit exactly into the mold of organic farming. Sustainable agriculture, with its emphasis on low inputs rather than no chemical inputs at all, sounded almost exactly like the New Agriculture that Louis Bromfield had promoted. As interest in Bromfield's farming methods revived in the late 1980s and 1990s, all of his farm books were reprinted in paperback by the Wooster Book Company, anthologies of Bromfield's best farm writings were compiled by Charles Little and George DeVault, Ohio University Press released a new edition of Ellen Bromfield Geld's *The Heritage*, and the Columbus PBS station, WOSU, even filmed two documentaries about Malabar Farm and Louis Bromfield. In 1988, Malabar Farm hosted its first workshop on regenerative agriculture, including "a wagon tour of the farm with emphasis on its history as a low-input and experimental farm under its noted founder, Louis Bromfield."[4]

It looked like Malabar was on its way back to becoming a landmark of sustainable agriculture. But park manager Brent Charette found that "the atmosphere in the park changed dramatically" in the late

1980s when "the state initiated collective bargaining." The state raised the salaries of its newly unionized workers but did not increase the operating budget for individual parks, leaving many parks, including Malabar, struggling to make ends meet. In 1989, frustrated by union politics, budget issues, and the fact that ODNR refused to make and follow a long-term management plan for Malabar, Charette decided to resign from state employment and "move into a family business," where he spent the rest of his career. The new park manager, Scott Doty, started at Malabar in July 1989. Doty had a degree from OSU in parks and recreation administration and had worked at various Ohio state parks—Tar Hollow, Scioto Trail, Quail Hollow, and Portage Lakes—for ten years before taking the position at Malabar.[5]

By 1991, ODNR's funding problems had become so severe that to balance the budget, the new administration, under Governor George Voinovich, laid off employees and curtailed programming. The positions of seventy-five ODNR employees, including two farm workers at Malabar, were terminated. One of the dairy workers who was laid off, Randy Swank, had worked at Malabar since 1972. Seven registered Hereford cows from the beef herd, some of them pregnant, were sold off in early November for far less than they were worth. The eleven-cow dairy herd was moved to OSU's Agricultural Technical Institute (ATI) in Wooster on November 12, even though the dairy barn at ATI was already full. Herd manager Stephen Hughes had to cull some of the cows he already had and slaughter two of the less productive Malabar cows—an Ayrshire and a Guernsey that had been purchased with money donated by the public—to fit in the remaining nine cows, which soon became integrated into the Wooster herd.[6]

Selling off the dairy herd was the least popular decision that the State of Ohio had made in its twenty-nine years of operating Malabar Farm. As soon as they heard about ODNR's decision to get rid of the dairy herd and lay off the farm workers, Malabar fans began to protest. Bromfield's daughter Ellen Bromfield Geld and granddaughter Melanie Read; former park managers Jim Berry and Brent Charette; Bromfield's first farm manager, Max Drake; OEFFA president Charles Frye; Malabar neighbor Annette McCormick; and many others were shocked when they heard the news and feared that it was ODNR's first step to eliminate farming at Malabar altogether. On November 23, over a hundred people attended a "Save the Cows" rally at Malabar in hopes that ODNR would change its mind and bring back the nine remaining dairy cows. In early 1992, Annette McCormick, her neighbor Judy FitzSimmons, and several other concerned citizens formed an action group called the "Malabar Farm Conservators" to pressure the state to return the cows to Malabar. On March 22, two hundred people braved snow and wind to show up for another rally and march down a road at Malabar with "Save the Cows" signs.[7]

Glen Alexander, the director of ODNR's Parks and Recreation division, tried to reassure concerned citizens that the state wasn't going to eliminate all farming operations at Malabar, just the dairy herd. Milking cows twice a day, seven days a week, cost the state $40,000 a year, and less than 1 percent of the total visitors to Malabar happened to wander by the dairy barn at 4:00 P.M. to watch the cows get milked. Since many people blamed park manager Scott Doty for selling the cows, ODNR moved him to a less controversial position, as assistant manager at Maumee Bay State Park. "Scott was made the fall guy for the whole thing," former park manager Brent Charette explained. "It was unfortunate what happened to Scott. He was too good a guy for that to happen to, in the wrong place at the wrong time." To replace Doty, ODNR appointed Louis Andres as the new park manager at Malabar. Andres had a degree in education and environmental science from Miami University in Oxford, Ohio, and had worked for ODNR at various state parks—Hueston Woods, Mohican, East Fork, and Cowan Lake—since 1980.[8]

Andres's first task as park manager was "damage control"—to counter the negative publicity and get people excited about what was happening at Malabar again. "The focus has been on the cows, but Malabar is much, much more," Andres wrote in June. "It's a historical landmark, a unique, working farm and a living classroom for educators and students to learn, enjoy and explore the land Louis Bromfield restored. . . . Let's step together into the future and let the past be our teacher." Andres saw Malabar's educational mission as twofold—to teach people about both Louis Bromfield and modern sustainable agriculture. During his first year as park manager, the Big House was repainted, the park purchased a new John Deere tractor and a bull for the beef herd, and several hundred people attended a series of hillside chats on Mount Jeez. In response to public demand for more agricultural education, the interior of the former dairy barn was converted to a museum containing antique farm machinery, a children's farm play area, and an egg incubator and chick brooder to show kids how chicks hatched and grew. With a grant from the George Gund Foundation, sustainable agriculture advocates cataloged and preserved the ecology library, which had sat forgotten and water-damaged in the Big House basement ever since the state took over Malabar. New titles donated by Rodale Press and others were added to the original 1950s books, and in 1993 the whole collection was moved to a small house by the working farm and opened to the public on a reference-only basis.[9]

As the dairy cow controversy highlighted, relying solely on state funding to operate Malabar risked entangling the farm in politics and budget cuts. Malabar was the only state park in Ohio that was "a working farm, historical site, and park all rolled into one," and it had unique expenses. Other state parks focused on building and maintaining campgrounds, trails, and restrooms; Malabar needed funds to buy tractors, preserve historic structures like the Big House, and eventually build a visitor education center. To help solve this problem, in 1990 park manager Scott Doty, ODNR leaders, and several others started a fundraising organization called "Friends of Malabar," under the umbrella of the Columbus Community Foundation. By 1992, Friends of Malabar had raised $70,000, which the group hoped to use "to form a significant base of capital, so interest earnings regularly could be drawn off to pay for construction and repairs at the park." While Friends of Malabar worked on fundraising, Louie Andres formed a ten-member advisory board in November 1992 "to supply technical assistance, guidance and support for development of the farm-park." In 1993, the advisory board merged with Friends of Malabar to form the nonprofit Malabar Farm Foundation, which worked over the years to both plan and fund historic preservation efforts and programming.[10]

By the end of 1992, things were definitely looking better at Malabar. Newspapers started giving the farm favorable publicity again, and the flood of angry letters about the cows slowed down to a trickle. Then came April 4, 1993. It was Palm Sunday, and Louie Andres was eating breakfast with his family when he heard sirens nearby. "I need to go down there and see what's going on," he said. When he arrived at Malabar, he was horrified to see that the main dairy barn was on fire. Earlier that morning, a park worker named Harry Taylor had entered the barn to feed the animals and discovered that the chicks were dead and there were flames "everywhere" around the brooder. The farm workers managed to get the twenty-seven Hereford cattle and most of the equipment out of the barn, but the fire had already reached the roof by the time the first fire trucks arrived. Despite the efforts of more than fifty firefighters from five volunteer fire departments, who pumped water out of the nearby pond and kept the flames from spreading to adjacent buildings, the big barn was completely destroyed, along with most of the antique farm machinery in the museum. It was afternoon before the fire was finally out and "the barn's remains had been bulldozed into a charred heap."[11]

No one knew exactly what had started the fire. The leading theory was that it was caused by the heat lamp igniting the bedding in the chick brooder. Another possibility was that a talking display, installed next to the brooder just a few days before the fire, had malfunctioned and overheated. Or perhaps the barn's electrical system had simply been overloaded by the high-wattage heat lamp in the brooder. Andres thought it might have been spontaneous combustion in stored hay. Over the next

The Malabar barn after the April 1993 fire. (Malabar Farm archives, courtesy of Ohio Department of Natural Resources)

two years, the fire was investigated by ODNR, the state fire marshal's office, the state highway patrol, and several other agencies, but they never did pinpoint an exact cause. All they concluded was that it was an accident, not arson, and it wasn't any particular person's fault.[12]

Even though he hadn't even been at Malabar when the fire started, many people blamed Andres for the loss of the barn, claiming that he and other employees had been incompetent and had ignored warnings that the chick brooder might be a fire hazard. "That was a very stressful year and a half; I got death threats," Andres said. "We were getting slammed in the media; it was nothing like the truth." He explained that ODNR's "way of solving the problem was to find a scapegoat, to find the person who caused the problem" and move him to another park, just like they had done a year earlier with Scott Doty. But Andres made up his mind that he was not going to be a scapegoat, and he had many friends who supported his work at Malabar and were willing to intercede for him. Influential people on the advisory board "made phone calls to the director and the governor." Finally, Andres recalled, "at ten o'clock one night I got a message from my boss, 'Call off your dogs, you won, you can stay.'"[13]

The first decision that ODNR had to make after the fire was whether it would build a replica of the original 1890s barn or replace it with a less expensive pole building. "It needs to be exactly the way it was," Andres insisted. As public outcries claiming that ODNR was mismanaging the farm increased again, Glen Alexander, chief of the Division of Parks, agreed to reconstruct the barn as an authentic timber-framed structure, despite the greater cost. Rudy Christian from the Timber Framers Guild volunteered his group's services to help design and participate in "the first public barn raising in North America for over a hundred years." Though no architectural drawings of the original barn existed, it had been the most-photographed structure in the park, and the architects discovered that the original 1890 barn had been well-built but the 1930s straw shed had been poorly constructed. The builders decided to properly frame both the main barn and the straw shed in the new barn. For the sake of historical authenticity, they relied on local timber as much as possible, using two hundred white oak trees that had been recently killed by gypsy moths in Mohican State Park to make beams for the new barn. Most of the preparatory work was done in Ohio, and some of the pieces for the barn were even sawed at the Malabar sawmill.[14]

Because the barn raising would be a historic event, the Timber Framers Guild and ODNR worked together to turn it into a three-day festival over Labor Day weekend in 1994. "A spectacular, but vanishing scene of Americana will be played out—perhaps for the last time on this scale—this weekend at Ohio's Malabar Farm State Park," ODNR wrote in its press release. A huge amount of planning went into the event. The general contractor, R. G. Beer, cleared the site and poured a new reinforced concrete foundation and basement to reduce the

Preparing the foundation for the new barn, 1994. (Malabar Farm archives, courtesy of Ohio Department of Natural Resources)

*Above:* Barn raising at Malabar, September 1994. (Malabar Farm archives; courtesy of Ohio Department of Natural Resources)

*Left:* Members of the Timber Framers Guild preparing to raise the barn. (Malabar Farm archives, courtesy of Ohio Department of Natural Resources)

risk of future fires. Preparing for thousands of visitors was a "logistics nightmare," Andres told reporters, with twenty extra park rangers and local police arriving to help direct traffic.[15]

Ten thousand people arrived on Saturday, September 3, to watch members of the Timber Framers Guild preassemble the major components of the barn and see steam engine enthusiasts demonstrate "bygone methods of sawing logs." An estimated thirty thousand people came on Sunday, set up chairs and blankets on Malabar's hills, and watched approximately 150 guild members raise the frame for the barn. When they weren't watching the actual construction, spectators wandered around and visited historical displays of steam engines and a "pioneer encampment," purchased food at "a small midway," bought copies of an oil painting of the barn raising by Mansfield artist Paul McClain, or carved their names on the wooden pegs that joined the barn together. The event concluded on Monday, when the contractor put on the walls and roof. In mid-September, workers hoisted three cupolas onto the roof of the barn, a local artist painted a mural on the door, and the new barn was complete. On the outside, it looked exactly like the original

*Right:* Raising the barn. (Malabar Farm archives, courtesy of Ohio Department of Natural Resources)

*Below:* The completed reconstructed barn, 1994. (Malabar Farm archives, courtesy of Ohio Department of Natural Resources).

barn, but it incorporated many safety features and was fully compliant with modern building codes.[16]

The new barn fit into a long-term plan that Andres first announced to the public in 1993: Malabar 2000. "Malabar Farm has the opportunity to become an international center for education and training in profitable, environmentally conscious, sustainable, grass-based farming; and to renew the devotion to the land started by Louis Bromfield over fifty years ago," a 1993 park brochure proclaimed. Under the Malabar 2000 plan, the gift shop and staff offices would be moved from the Big House into a new visitor center building, allowing the Big House and other historic structures to be restored to look like they did in Bromfield's day. Meanwhile, the working farm and fields would be used to educate visitors about modern sustainable agriculture "as it is and should be," not "nostalgically as it was, as historical farms do." Representatives from OEFFA and OSU's sustainable agriculture program hoped to make Malabar a showplace for conservation tillage and *management intensive grazing,* a system of pasture management that helped keep both cattle and pastures in optimum condition by frequently moving the animals to a new paddock. Ohio Agricultural Research and Development Center (OARDC) dairy scientist David Zartman was experimenting with a combination of management intensive grazing and seasonal dairying on a farm in northeastern Ohio and suggested implementing a similar system at Malabar.[17]

Though the state still couldn't afford to pay workers to milk a dairy herd, it did at least partially switch to management intensive grazing with the beef herd. In July 1994, the park held "its first intensive

Pastures and strip cropping at Malabar, 1990s. (Malabar Farm archives, courtesy of Ohio Department of Natural Resources)

grazing management field day." A sign explaining the "grazing cell" at Malabar was erected in one of the pastures. Even the gift shop got into the grazing theme, selling "cow pies"—which were actually "hand-sculpted, 1 lb. chocolate droppings of fudge," packaged in a box lined with artificial "grass." But as a state agency, Malabar was limited in just how far it could go toward producing completely grass-fed beef. The Ohio Department of Agriculture "wanted to make sure we didn't make Malabar go organic," Louie Andres recalled. One ODA representative even warned him to make sure that no beef animal left Malabar "without having grain in it." Partly because of funding difficulties but more because of a strong antiorganic attitude at the state level, Andres was never able to fully implement all of the sustainable agriculture practices recommended by the Malabar Farm Foundation. Disappointed at this compromise, critics complained in the local media that "the state isn't keeping the promises it made to farm Malabar."[18]

Sign describing grazing cells at Malabar, 1990s. (Malabar Farm archives, courtesy of Ohio Department of Natural Resources)

The negative publicity about Malabar was part of a larger critique claiming that ODNR was "failing" at protecting and preserving Ohio's natural resources; other local complaints included ODNR's plan to selectively harvest trees in the Mohican State Forest and concern about a landfill being constructed near the Fowler Woods Nature Preserve. Like other state and federal conservation agencies in the 1980s and 1990s, ODNR was caught in the middle of intense culture wars about natural resource management. By 1995, a large percentage of environmentalists had adopted a biocentric worldview, which held that nonhuman species and organisms had the same inherent value and right to existence as human beings. Expanding on the philosophies of some of Bromfield's contemporaries, like Paul Sears and Aldo Leopold, biocentrists in the 1980s and 1990s sought to preserve as much wilderness—natural areas unmarred by humans—as possible, where species could continue to live and die and evolve as they had for billions of years before humans disrupted the natural environment.[19]

Resource users of all persuasions, from profit-oriented lumber companies to traditional conservation agencies that advocated sustainable forestry, felt threatened by the suggestion of some of the more radical environmentalists that all logging and grazing on public lands be halted immediately. Some early antienvironmental groups, like the "Sagebrush Rebellion" of the late 1970s, wanted to transfer all federal lands to private ownership. Other free-market environmentalists had a more nuanced perspective—they endorsed federal wilderness protection of some areas for aesthetic and recreational reasons but thought that a market-based policy would be the best way to manage renewable natural resources and abate pollution. Like the Progressive and New Deal conservationists, these "counterenvironmentalists" were unashamedly anthropocentric and argued that it was both possible and necessary to sustainably manage renewable natural resources for human use. Criticism of the emphasis on wilderness as an idealized nonhuman landscape was also raised by environmental historians like William Cronon, who pointed out that practically every landscape on earth had been at least minimally influenced by humans.[20]

Traditional conservation agencies like ODNR were caught in the middle of these debates. Once perceived as good stewards of public natural resources, they were now accused by biocentric environmentalists as being sold out to industry if they allowed logging in state forests or drilling for natural gas on public lands. Similarly, the Ohio Department of Agriculture—which still had the final say over which agricultural practices were used at Malabar Farm—felt threatened by the growing organic movement and emphasized that conventional agriculture, if done right, was sustainable. Since Malabar was both a farm and a state park, it was simultaneously caught in the public land management and sustainable agriculture debates. "We had to be careful; we would say we were just demonstrating a lot of principles of agriculture, organic as an option," Louie Andres explained. "They were okay with that as long as we had traditional agriculture and as long as we weren't going off into wild stuff. That was a fine line we had to walk."[21]

Andres was frustrated that the local media failed to appreciate his attempt to take a middle-of-the-road stance on contentious issues. After seeing himself misquoted in yet another negative article in the *Mansfield News Journal,* he got mad and called the reporter, Ron Simon, to complain. On the other end of the line, he heard nothing but a "dead silence." After a while, someone else picked up the phone and told Andres that Simon had just had a heart attack. "I felt really bad," Andres said. "I probably killed this guy!" Fortunately, Simon recovered, and he called Andres back a couple weeks later. After Andres apologized for getting angry, Simon explained that he was hard of hearing and that his editors had changed some things in the article. The two men became friends after that, and in October 1995 Simon wrote an article commending Andres's efforts to continue Bromfield's legacy at Malabar. The problem, Simon realized, was that "Louis Bromfield was the golden age of Malabar," and "we can't bring that back any more than we can bring Bromfield back." For better or worse, times were changing, and with its triple function of historical site, working farm, and recreational area, Malabar had to change as well.[22]

Though the state would not allow the farm to pursue official organic certification, Louis Andres partnered with OEFFA in 1999 to once again plant an organic vegetable garden at Malabar. He worked with Charlie Frye at OEFFA and Ben Stinner at OARDC to come up with a plan, interviewed several people, and hired Trish Mumme, a "world religion professor from Granville," and two interns to live in the Ceely Rose house and run the garden. Mumme and her interns planted a variety of vegetables, including tomatoes, sweet corn, squash, peas, beets, bok choi, kohlrabi, Swiss chard, and beans. They sold the vegetables at the roadside stand and also started a CSA (community supported agriculture) program where shareholders picked up a bag of fresh vegetables once a week from June to October. The contract with Mumme was originally for five years, but the partnership ended after three years because of personality conflicts and other difficulties.[23]

As books like Michael Pollan's *The Omnivore's Dilemma* drew attention to a growing local food movement in the early 2000s, Andres reoriented the Malabar Inn's menu from typical American food (often purchased readymade from a food service supplier and heated in the microwave) to French cuisine made with as many local ingredients as possible, including Malabar-raised beef and local vegetables. In 2006, the name was changed to the Malabar Restaurant, and in 2007 the park leased the restaurant to chef Daniel Bailey. With a new and improved menu and an emphasis on fresh, local food, attendance at the restaurant quickly increased. Andres said he used to be able to "sit at any table I wanted; now I had to make reservations, it got so crowded."[24]

To solve the ongoing problem of insufficient farm labor due to budget cuts and unionization, Malabar Farm entered into a partnership

The Malabar Farm Restaurant began using local ingredients in the 2000s. (Photo by author)

with the Mansfield Correctional Institution (MANCI), which operated its own prison farm. MANCI brought labor, equipment, and cattle to Malabar Farm during the summer and then butchered beef animals in the fall, splitting the produce fifty-fifty with ODNR. While Andres and other park staff believed bringing in prison labor was an ideal solution to their perennial labor problems, several local residents worried about the safety of having prisoners working at the farm. A MANCI representative assured them that "inmates on work detail can't have sex-related offences or attempted escapes on their record," and there were no incidents during all the years prisoners worked at Malabar.[25]

When the year 2000 arrived, the centerpiece of the Malabar 2000 plan—a new visitor center—had not yet been started. At the time, no other Ohio state parks had visitor centers, and ODNR did not want to budget the money to build one at Malabar. But the need was acute—having the staff offices and gift shop in the Big House was putting too much wear and tear on the historic structure. The state installed a new geothermal heating and cooling system in 2002 to help preserve the artifacts in the house by keeping the temperature constant year-round; moving the staff offices to a different building was the next step for historical preservation. Finally, Andres convinced Glen Alexander to agree to a matching fund—if private donors would put up half the money to build a visitor center, ODNR would provide the rest. The Malabar Farm Foundation spearheaded fundraising efforts, holding annual "MalaBar-BQs" to raise money for the new visitor center. One of the first donors was Ralph Cobey, who gave $50,000 toward the project—more than making up for the $26,000 he had charged the state to buy Malabar back in 1972.[26]

The gift shop was still in the Big House garage in the 1990s. (Malabar Farm archives, courtesy of Ohio Department of Natural Resources)

Staff office in the Big House, 1990s. (Malabar Farm archives, courtesy of Ohio Department of Natural Resources)

By 2005, the Malabar Farm Foundation had raised enough money for construction on the visitor center to begin. The State Controlling Board released $900,000 for the project and the foundation provided $500,000 to build the $1.4 million structure. At a formal groundbreaking ceremony during Heritage Days on September 24, 2005, Governor Robert Taft, Ellen Bromfield Geld, Louis Andres, and twelve other people "turned a symbolic bit of sod with a gold-plated shovel." The visitor center was constructed near the 1980 parking lot, in Bromfield's former dairy pasture a short distance from the Big House. In line both with Bromfield's original vision and heightened contemporary interest in sustainability, Louie Andres insisted that the visitor center incorporate recycled materials, be well-insulated and energy efficient, and demonstrate the use of renewable energy. "We are supposed to be green, sustainable; let this be an example," he said. "We have to walk the talk." A one-kilowatt wind turbine was installed on a hundred-foot-high tower, along with some solar panels and a battery back-up system.[27]

The new visitor center was completed in 2006. (Photo by author)

Despite its unconventional design, construction of the visitor center kept on schedule, and the building was complete in time for its grand opening on September 23, 2006, during the annual Heritage Days celebration. Taft and Geld, along with Andres and Malabar Farm Foundation president Ellen Haring, "cut a little ribbon to officially open the farm's Visitor Education Center." The new building contained new and larger staff offices, fifteen themed educational exhibits, and a new library / conference room to house most of the books from the sustainable agriculture library. The expanded gift shop offered Malabar-raised beef, eggs, maple syrup, reprints of Bromfield's books, clothing, fudge, and many other items. From September 2006 on, the new visitor center was the initial point of contact between park guests and staff, where visitors could say hello to Poncho the parrot, visit the "live songbird aviary" behind the building, buy house- and farm-tour tickets, and embark on wagon tours of the farm.[28]

During the 2000s, Malabar continued to offer a wide range of programs for visitors of all ages. Heritage Days, the Maple Syrup Festival, Spring Plowing Days, and hearthside cooking classes remained annual traditions at Malabar. Other programs changed with the times; at the kid-friendly Farm Olympics, children who participated in the "tobacco spitting" contest chewed and spit Tootsie Rolls instead of real tobacco. Starting in 2003, the park hosted a series of Bromfield-inspired dramas by local playwright Mark Sebastian Jordan, performed in the new barn. In August 2005, the park capitalized on "the recent media interest in ghost hunting" with a Night Haunt, where visitors could take a three-mile hike in the dark to look at bats and go on ghost tours of the Ceely

Rose house and the Big House. Even as interest in the darker side of supernatural phenomena increased, a group of atheists criticized Malabar in 2007 for displaying a nativity scene as part of its annual Christmas celebration. The state's compromise, implemented in 2008, was to put up the nativity scene but accompany it with a "secular display," including a Christmas tree.[29]

In 2008, Louie Andres was promoted to a regional park management position, and assistant manager Jason Wesley took over operations at Malabar until November 2012, when ODNR hired a new park manager, Korre Boyer. Boyer had a degree in agricultural education from OSU, raised beef cattle and grain on his own two-hundred–acre farm, and was the first Malabar park manager that ODNR had hired from outside the agency. Boyer planned to make Malabar a showplace for sustainable agriculture, as defined by OSU and organizations like the Ohio Farm Bureau Federation. He had nothing against organic farming in principle, but he didn't believe it was a viable way to feed the world. Instead, he emphasized modern conventional practices like *no-till* corn and soybean production, which used broad-spectrum herbicides and genetically engineered herbicide-resistant crop plants to eliminate tillage altogether. Many sustainable agriculture advocates, including OSU soil science professor Rattan Lal, promoted no-till as the best possible cropping system to simultaneously halt erosion and combat climate change by sequestering carbon in the soil. Boyer kept the fields in strip-

Strip cropping of hay and no-till corn at Malabar, 2015. (Photo by author)

Repainted barn and mural, 2018. (Photo by author)

cropping as much as possible, alternating strips of hay with corn and soybeans. He hoped to eventually offer more educational programs at Malabar but thought he needed to get the fields looking good first.[30]

In addition to using the most up-to-date farming practices at Malabar, Boyer oversaw critically needed exterior renovations to the Big House. In 2013, the state released $500,000 to strip fourteen layers of paint off the Big House and repaint it, along with refinishing several other buildings. The big barn was repainted, and several local retired art teachers—Tom Bachelder, Mark Summer, and Kathy Lowe—carefully traced and painted a replica of Louis Bromfield's original barn mural on the barn door. Along with these renovations, the Malabar Farm Foundation introduced several new park events, including a romantic Valentine's dinner for couples featuring a Frank Sinatra impersonator, "Malabark in the Park" for dog-lovers and families, monthly barn dances, and a "Sunday Drive Car Show" featuring classic automobiles. The Malabar Restaurant, operated by Joey Motter and Kim Williams for most of the decade, continued to serve high-quality meals featuring Malabar beef, local vegetables, and homemade pasta and bread.[31]

Local environmentalists and organic farming advocates criticized Korre Boyer's conventional agriculture mindset, which may have influenced his decision to resign from his position as park manager in January 2018. The Malabar Restaurant and youth hostel also closed that year, and the farm's fields were leased to a neighboring farmer who did not use very good management practices. It was May 2019 before the park hired a new manager, Jennifer Roar. Roar had a master's degree in environment and natural resources from OSU and had worked for several years as an educator for ODNR's Division of Watercraft. She spent most of her first year as park manager doing cleanup work, including removing out-of-control multiflora rose and renovating the Malabar Restaurant and other buildings. Then came the COVID-19 pandemic, and beginning in March 2020 the Big House and visitor center at Malabar were closed to visitors. The park was able to hold the first weekend of the Maple Syrup Festival as scheduled, but the second weekend and all other events in 2020—including Heritage Days—had to be canceled. Only the park's hiking trails and outdoor areas remained open during the pandemic, but Roar was able to "make the best of that situation" by focusing on maintenance and cleanup work that there might not have been time for otherwise.[32]

Despite the challenges caused by the pandemic, Roar was excited about Malabar's future when interviewed in July 2020. "There is a balance between keeping the history and the legacy and implementing what is ideal here as far as innovation and education," she said. "How do we implement these new innovative techniques for today but still remember to talk about the history?" Programming at Malabar has always tried to balance these two goals—preserving the past while keeping up with the present. It's a sometimes contradictory mission, and much of the criticism of Malabar's management over the years has stemmed from outside observers not understanding how difficult it is to maintain this balance. In an increasingly polarized world, it is challenging to take a middle-of-the-road position like Louis Bromfield did. Yet finding that middle ground is more important than ever.

# Conclusion

Much has changed in the world—and at Malabar Farm—since that long-ago wintery day when Louis Bromfield knocked on the door of an old Ohio farmhouse. In 1939, nearly a quarter of Americans lived on farms. Tractors were just beginning to replace horses in most Ohio fields, dairy cows spent their summers on pasture, and the milk that was delivered in glass bottles to the doorsteps of Mansfield homes came from small farms in the local milkshed. Yet the reality of American farming in the 1930s was not as idyllic as Bromfield had dreamed. Debt, drought, and severe soil erosion plagued a high percentage of American farms, threatening the livelihoods not only of farmers but of the nation as a whole. "We all live on, or from, the soil," Friends of the Land warned in their manifesto. "No matter which political party gains ascendancy as the years go by; whether the swing be from middle Left to far Right, or to the farther Left; whether we remain at peace or go to war again, this fact will remain: so long as we keep on scrubbing off, blowing off, killing off our topsoil, business and social conditions in this country will remain fundamentally unsound."[1]

Today, less than 1 percent of the US population grows food for everyone else—a development that has been both hailed as a triumph of efficient agribusiness and decried as the loss of a rural way of life. The American food system has become increasingly complex and opaque, and many people neither know nor care where the food they eat comes from beyond the grocery store or restaurant where they purchased it. They may not realize that milk comes from cows, and even if they do, it's impossible to figure out which cow on which farm produced the milk in any given plastic gallon jug at the grocery store. The origins of processed foods are even more obscure. Few people could even name all of the raw materials that went into a frozen pizza or a chicken nugget, let alone trace them back to the farms where they were grown. Yet as nameless and faceless as the modern American food supply may seem, the fact remains that we are just as dependent on the soil for our lives today as people were in the 1930s.[2]

The majority of American farmers understand, and are proud of, their role in feeding the United States and the world. They view food production as a patriotic duty and become defensive when anyone (especially a nonfarmer) criticizes their production practices. After all, they are the ones who survived the massive farm consolidations, who embraced the concept of agribusiness, and who outcompeted their neighbors by being the first to adopt new and improved farming techniques. In one sense, modern agribusiness can be viewed as a direct descendant of the New Agriculture Louis Bromfield promoted so passionately in the 1940s and early 1950s. For twenty years after his death, the managers of Malabar Farm believed making the farm into a showplace modern dairy operation was the best way to honor Bromfield's legacy; more recently, park manager Korre Boyer operated the farm using similar logic.

But modern conventional agriculture, despite its success in producing ever-rising yields with ever-decreasing labor, has had its share of unintended environmental consequences. Nitrogen and phosphate fertilizers, when used in excess, caused algal blooms in lakes and polluted wells. Pesticides like DDT biomagnified in the food chain and killed robins and raptors; newer pesticides are less well-studied but still toxic to nontarget organisms. Even the herbicide glyphosate, which is the keystone of modern no-till farming, is coming under increasing scrutiny as a possible human carcinogen. And the increased output per laborer in modern agriculture is only possible with the use of ever-larger tractors and other gasoline or diesel-burning farm machinery. In the 1970s, the vulnerability of mechanized agriculture to disruptions in oil supplies became alarmingly obvious; today, environmentalists are more concerned about the contribution greenhouse gases from agriculture may make to climate change. When Malabar became a state park in 1976, environmentalists emphasized Bromfield's original self-sufficient Plan for Malabar, his early warnings about the dangers of pesticides, and the horse-drawn, diversified agriculture he reminisced about in his novel *The Farm* but never actually practiced at Malabar.

From the 1980s on, conventional farmers and environmentalists have often been in conflict with each other. The phrase *sustainable agriculture* was coined in an attempt to make natural farming methods palatable to farmers who were turned off by the word *organic*. In many ways, the emphasis on sustainability improved agricultural practices, but the word was soon "co-opted" by agribusiness and used in reference to any new agricultural technology, from no-till farming using herbicide-resistant genetically engineered crops to precision agriculture that uses GPS and information technology to strategically apply fertilizers and pesticides. This debate about what sustainable agriculture should look like was at the root of many of the controversies over farming practices at Malabar in the 1990s and, to some extent, continues today.[3]

Over the years, specific aspects of Bromfield's legacy have been highlighted by various groups who have radically different ideas about what the future of agriculture should look like. What has rarely been emphasized, however, is that *all* of these concepts—agribusiness, self-sufficiency, sustainability—are integral to Malabar Farm. To see Malabar strictly as a modern production farm, or a nature preserve, or the home of a famous novelist oversimplifies the complexity of what Louis Bromfield actually did. Malabar wasn't a conventional farm or an organic farm; it was *both*. It represents a middle ground that is often lacking in modern discussions about sustainability or environmental issues, yet remains critically important. Too often the problems of sustainable agriculture are portrayed in black or white, yes or no terms: organic farming or chemical-saturated GMOs; moldboard plowing or no-till; grass-fed or feedlot beef; homesteads or mega-farms. Emotional, passionate advocates of both conventional agriculture and environmental protection rarely attempt to understand where their opponents are coming from, leading to misunderstandings, hurt feelings, and angry reactions on both sides.

As Louis Bromfield and Friends of the Land realized, the use of natural resources is essential to human survival. Between the extremes of unchecked exploitation and complete nonuse, they advocated conservation—the wise use, or stewardship, of all natural resources, especially soil. Conservation, as they defined it, required that all natural resource management, including farming, be based on an understanding of and respect for natural processes. "No man ever yet won a victory by *fighting* nature and the laws of nature," Louis Bromfield warned. "By working with nature man can be prosperous and even rich and happy and healthy. Fighting or cheating her, man is always defeated, poverty-stricken, bitter and miserable, and eventually is destroyed himself." To Louis Bromfield, Friends of the Land, and other soil conservationists in the 1940s, working with nature was the only way to conserve soil and create a permanent agriculture.[4]

As scientists, farmers, and environmentalists discuss how to make American agriculture more sustainable, the history of Malabar Farm, and the soil conservation movement in general, can give a much-needed historical perspective on what has worked in the past and what has not. Current advocates of sustainable agriculture can learn just as much from the mistakes and failed dreams of New Deal conservationists as they can from their successes. For example, even the best-designed, most environmentally friendly agricultural systems cannot by themselves eliminate the broader social and economic problems of society. If they could, then those problems would have been resolved seventy years ago by the New Deal conservationists. Ironically, some of the solutions advocated in the 1940s ended up causing problems we are still dealing with today—like nutrient pollution in waterways from excessive fertilizer runoff, or invasive plants like multiflora rose and kudzu that were once promoted as soil-conserving crops, or the excessive growth of industrialized agribusiness to the point where less than 1 percent of the American population grows the food for everyone else.

While it is true that good agricultural practices alone cannot solve all of society's problems, it is equally true that a sustainable agricultural system is a prerequisite to any successful civilization. From an ecological and environmental perspective, the soil conservation and grass farming practices advocated by Louis Bromfield seem sustainable. As recent interest in management intensive grazing and organic farming shows, advances in scientific research since Bromfield's day can make these practices even more productive than they were seventy years ago. Then as now, an environmentally sustainable system of farming is within reach. The very fact that many of the ideas of the soil conservation movement have been revived in recent years, despite what seemed almost overwhelming odds, considering how they were marginalized in the 1950s, demonstrates their timeless value. Somehow, someway, the concept of working with nature to grow food will endure, no matter which word is eventually used to describe it.

It is likely that the future of sustainable agriculture, as well as future programming at Malabar Farm, will be a mixture of old and new. We cannot cut and paste Bromfield's farming methods directly into the modern agricultural system. They were a product of their time and changed even during Bromfield's own lifetime—from a self-sufficient "island of security" to a specialized grass farm to a diversified operation selling direct-market pesticide-free vegetables and pastured pork. Today, his Ferguson tractors are collector's items, the small local milksheds are gone, and the health department frowns on putting vegetables directly in "non-potable" spring water. But all the changes in agriculture since the 1940s are cultural constructs; the land remains the same. The fundamental principles advocated by Bromfield and the soil conservationists are still valid—healthy soil is the foundation of a successful agriculture, urban people should care even more about soil health than farmers, and the only way to have a permanently successful farming system is to work with nature instead of against it. Perhaps Louis Bromfield himself best summed up the true legacy of Malabar Farm, as he reflected back upon his original vision for Malabar in his final book, *From My Experience:*

> A return to the past can never be accomplished and the sense of fortified isolation and security is no longer possible in the world of automobiles, of radios, of telephones and airplanes. One must live with one's times and those who understand this and make the proper adjustments and concessions and compromises are the happy ones. In the end I did not find at all what I was seeking on that snowy night; I found something much better . . . a whole new life, a useful life, and one in which I have been able to make a contribution which may not be forgotten overnight and with the first funeral wreath, like most of the writing of our day, but one which will go on and on. And I managed to find and create, not the unreal almost fictional life for which I hoped, but a tangible world of great and insistent reality, made up of such things as houses, and ponds, fertile soils, a beautiful and rich landscape, and the friendship and perhaps the respect of my fellow men and fellow farmers.[5]

Malabar Farm is that tangible legacy of Louis Bromfield. May it always continue to celebrate both the past and the present, preserving Bromfield's history and emphasizing farming practices that work with nature.

# Notes

## Abbreviations

*ABJ* *Akron (OH) Beacon Journal*
ATI Agricultural Technical Institute, Wooster, Ohio
*BTF* *Bucyrus (OH) Telegraph-Forum*
CCC Civilian Conservation Corps
*CD* *Columbus (OH) Dispatch*
*CPD* *Cleveland (OH) Plain Dealer*
FOTL Friends of the Land Records Papers, Ohio History Connection
IWLA Izaak Walton League of America Records (Ohio), Ohio History Connection
LBC Louis Bromfield Collection, Ohio State University
*LW* *Land and Water*
MF Malabar Farm archives, Malabar Farm
*MFN* *Malabar Farm Newsletter*
*MNJ* *Mansfield (OH) News Journal*
*MS* *Marion (OH) Star*
NF Samuel Roberts Noble Foundation Archives, Noble Research Institute
NRCS Natural Resources Conservation Service
OAES Ohio Agricultural Experiment Station, Wooster
OARDC Ohio Agricultural Research and Development Center, Wooster (formerly OAES)
ODA Ohio Department of Agriculture
ODNR Ohio Department of Natural Resources
OEFFA Ohio Ecological Food and Farm Association
OSU Ohio State University
SCS Soil Conservation Service
SP Sears (Paul Bigelow) Papers, Yale University
TVA Tennessee Valley Authority
USDA US Department of Agriculture
*ZTR* *Zanesville (OH) Times-Recorder*

## Introduction

1. "Malabar Farm: The Most Famous Farm in the World," Friends of the Land promotional flyer, box 11.2, FOTL.

2. This is not to say that these agricultural histories were not valuable, as they did highlight many important factors influencing the changes in American agriculture. For primarily economic summaries of twentieth-century agricultural history, see Cochrane, *American Agriculture;* Gardner, *American Agriculture;* and Schlebecker, *Whereby We Thrive.* The policies that have influenced agriculture are summarized by Fite, *American Agriculture* and Hurt, *Problems of Plenty.* Histories of technological developments include Joseph Anderson, *Industrializing the Corn Belt;* Conkin, *Revolution Down on the Farm;* Drache, *U.S. Agriculture;* Fitzgerald, *Every Farm a Factory;* Hart, *Changing Scale;* Holbrook, *Machines of Plenty;* and Olmstead and Rhode, *Creating Abundance.*

3. A. Dan Tarlock, "Rediscovering the New Deal's Environmental Legacy," in Henderson and Woolner, *FDR and the Environment,* 156–57; Cronon, "Place for Stories"; Nash, *Wilderness,* 380. For examples of this declensionist environmental version of history, see Adams and Newhall, *American Earth;* Fleming, "Roots of the New Conservation Movement"; Kline, *First Along the River;* and Shabecoff, *Fierce Green Fire.* Most of these authors followed the lead of Samuel Hays, who highlighted Progressive conservation and the 1970s environmental movement but not New Deal conservation in his books *Conservation and the Gospel of Efficiency* and *Beauty, Health, and Permanence.*

4. William Cronon, "The Trouble with Wilderness; or, Getting Back to the Wrong Nature," in Cronon, *Uncommon Ground,* 85; Worster, "Transformations of the Earth"; Cronon, "Modes of Prophecy"; Sutter, "World with Us," 94–96, 104–9; Igler, "On Vital Areas," 120; Steinberg, "Down to Earth," 803; Ruuskanen and Väyrynen, "Theory and Prospects," 462–65; McNeill, "Observations," 34–39; Hersey and Vetter, "Shared Ground," 413–31. For examples of environmental history on New Deal soil conservation, see Worster, *Dust Bowl;* Hurt, *Dust Bowl;* Beeman and Pritchard, *Green and Permanent Land;* Henderson and Woolner, *FDR and the Environment;* and Sutter, *Famous Gullies.*

5. USDA, *2015 National Resources Inventory,* 2–1; Witt, *Rural Sociology,* 8; US Bureau of Labor Statistics, "Labor Force Statistics from Current Population Survey," 2016, https://www.bls.gov/cps/cpsaat01.htm; Ian D. Wyatt and Daniel E. Hecker, "Occupational Changes During the 20th Century," *Monthly Labor Review,* Mar. 2006, 37–57; Stella Fayer, "Agriculture: Occupational Employment and Wages," *Monthly Labor Review,* July 2014, https://doi.org/10.21916/mlr.2014.28.

6. "Friends of the Land—Our Purposes and Program," *LW* 2, no. 3 (1956): 9–10.

## 1. Louis Bromfield Comes Home

1. Bromfield, *Pleasant Valley,* 3, 13–14; Bromfield, *From My Experience,* 6.

2. Bromfield, *Farm,* 291–94, 306–7; Scott, *Louis Bromfield,* 3. For more details on Bromfield's childhood and family history, see his fictionalized autobiography, *The Farm.*

3. Bromfield, *Farm,* 90–94, 295–305, 324–44; Bromfield, *Pleasant Valley,* 53–54; David Anderson, *Louis Bromfield,* 24.

4. David Anderson, *Louis Bromfield,* 24, 46. For more on Bromfield's life and literary career, see also Morrison Brown, *Louis Bromfield;* Heyman, *Planter of Modern Life;* Geld, *Heritage;* and Scott, *Louis Bromfield.*

5. Scott, *Louis Bromfield,* 138; Bromfield, *Pleasant Valley,* 8–9, 68–69; Geld, *Heritage,* 1–67.

6. Bromfield, *Pleasant Valley,* 21–22. For the pre-Bromfield history of Malabar Farm, see Bachelder, *These Thousand Acres.* Bromfield fictionalized the names of the former property owners (except Ferguson) in his books: he called the Herring farm the "Anson place"; the Beck farm the "Fleming place"; and the Niman farm (also known as the Schrack farm) the "Bailey place."

7. Bromfield, *Pleasant Valley,* 69–89; Heyman, *Planter of Modern Life,* 157–68; Ruth Smith, "Architect Lonesome for 'Old' Malabar," *MNJ,* Jan. 7, 1968.

8. Bromfield, *Pleasant Valley,* 62–72; Bromfield, *From My Experience,* 3–8.

9. Bromfield, *From My Experience,* 7–8, 209; Bromfield, *Malabar Farm,* 150; Bromfield, *Out of the Earth,* 145; Bromfield, *Pleasant Valley,* 50; Max Drake, "Max Drake Remembers Bromfield," in DeVault, *Return to Pleasant Valley,* 161; D. R. Dodd, *Erosion Control,* 36–40; Conrey, Cutler, and Paschall, *Soil Erosion in Ohio,* 10.

10. Cronon, *Changes in the Land,* 12, 22–57; Petulla, *American Environmental History,* 21–22; Udall, *Quiet Crisis,* 54; Krech, *Ecological Indian,* 99–122; Chad Anderson, "Native North America"; Munoz et al., "Spatial Patterns"; Mt. Pleasant, "Paradox of Plows."

11. Judd, *Untilled Garden,* 219; Lord, *To Hold This Soil,* 17–18; Sears, *Deserts on the March,* 44.

12. Judd, *Untilled Garden,* 240–43; Hugh Bennett, *Soil Conservation,* 868–71; McDonald, *Early American Soil Conservationists,* 2, 4–6, 31; Banner, *How the Indians Lost Their Land.*

13. Wulf, *Founding Gardeners,* 10, 85, 104, 116; Hugh Bennett, *Thomas Jefferson;* McDonald, *Early American Soil Conservationists;* Hugh Bennett, *Soil Conservation,* 868–98; Stoll, *Larding the Lean Earth,* 47, 52–53, 106, 168; Swanson, *Golden Weed,* x, 54–59, 80–117, 196–224, 255.

14. Judd, *Common Lands,* 95; Judd, *Untilled Garden,* 276; Stoll, *Larding the Lean Earth,* 175–81; Marsh, *Human Action,* 33, 39, 43–44, 329–30, 386.

15. Pinchot, *Breaking New Ground,* 505; David Stradling, "Introduction," in Stradling, *Progressive Era,* 8–12; Henry Clepper, "The Conservation Movement: Birth and Infancy," in Clepper, *Origins of American Conservation,* 3–15; Beatty, "Conservation Movement," 10; Tyrell, *Wasteful Nation,* 9, 81–82, 87. For more on Progressive conservation, see Frank Graham Jr., *Man's Dominion* and Samuel Hays, *Gospel of Efficiency.*

16. Pinchot, *Breaking New Ground,* 3–10, 23–29, 40–44, 102, 135–37, 190, 256. For more on Gifford Pinchot, see Pinkett, *Gifford Pinchot;* Miller, *Making of Modern Environmentalism;* Miller, *American Conservationist;* and Samuel Hays, *Gospel of Efficiency.*

17. Bailey, *Liberty Hyde Bailey,* 99; Shulman, "Business of Soil Fertility," 410–12; Tyrell, *Wasteful Nation,* 132–33; Pinchot, *Fight for Conservation,* 9; Hopkins, *Soil Fertility,* 183, 199; Tanner and Simonson, "Franklin Hiram King"; Franklin King, *Farmers of Forty Centuries;* Heckman, "Soil Fertility Management," 2796–98; Hersey, *My Work;* Carver, *Worn Out Soils.*

18. Whitney, *Soils of the United States,* 66; Hopkins, *Soil Fertility,* 340–41; Fanning and Fanning, "Milton Whitney"; Helms, "Early Leaders"; Sutter, *Famous*

*Gullies,* 40–60; Amundson, "Philosophical Developments"; Cochrane, *American Agriculture,* 99–100; Conkin, *Revolution Down on the Farm,* 27; Fite, *American Agriculture,* 4; Gardner, *American Agriculture,* 1; Schlebecker, *Whereby We Thrive,* 151; Hurt, *Problems of Plenty,* 12–13.

19. Brink, *Big Hugh,* 56–57; Hugh Bennett, *Soil Conservation,* 96–97; Blanco and Lal, *Soil Conservation,* 21–25; Hugh Bennett, *Soil Erosion,* 6.

20. Lord, *To Hold This Soil,* 49; Chase, *Rich Land,* 94–99; Hugh Bennett, *Soil Conservation,* 510; Sutter, *Famous Gullies,* 166–79. See Sutter, *Famous Gullies,* for a detailed discussion of Providence Canyon.

21. Ross Calvin, "Goodbye, Main Street: The History of a Gully," *Land* 2, no. 2 (1942): 121–24; Conrey, Cutler, and Paschall, *Soil Erosion in Ohio,* 3.

22. Hugh Bennett, *Soil Erosion,* 5; Helms, *Erosion Service,* 2; Brink, *Big Hugh,* 74–79; Sampson, *Love of the Land,* 4.

23. Duncan, Burns, and Dunfey, *Dust Bowl,* 17, 35; Hurt, *Dust Bowl,* 19–27; Helms, *Readings,* 152–53; Worster, *Dust Bowl,* 80–100.

24. Duncan, Burns, and Dunfey, *Dust Bowl,* 56, 64, 80–86, 100, 183; Hurt, *Dust Bowl,* 34, 53; Worster, *Dust Bowl,* 13–29.

25. Helms, *Readings,* 6, 137; Brink, *Big Hugh,* 6–7; Sampson, *Love of the Land,* 8–13; Helms, *Creation of the Soil Conservation Service,* 18–21.

26. Hugh Bennett, *Soil Conservation,* 319–21; Headley, "Soil Conservation," 189; Sears, *Deserts on the March,* 44, 144. For more on the connection between ecology and soil conservation, see Clements, *Dynamics of Vegetation;* Clements, "Experimental Ecology"; Thornthwaite, "Research Program," 218–20, 236; Whitfield, "Ecological Relations," 262–70; Lowdermilk, "Ecological Principles," 54, 99; Edward Graham, "Ecology and Land Use," 123–28; and Tobey, *Saving the Prairies.*

27. Hugh H. Bennett and Walter C. Lowdermilk, "General Aspects of the Soil-Erosion Problem," C. R. Enlow and G. W. Musgrave, "Grass and Other Thick-Growing Vegetation in Erosion Control," and Walter V. Kell, "Strip Cropping," all in USDA, *Soils and Men,* 602, 619–20, 624, 634–45; Hugh Bennett, *Soil Conservation,* 346–63; Blanco and Lal, *Soil Conservation,* 180–83.

28. Scott, *Louis Bromfield,* 319–20, 333–34, 379–81; Geld, *Strangers in the Valley,* 3–4; Heyman, *Planter of Modern Life,* 176–80; "Max Drake Remembers Bromfield," 160–61; "Blair Maxwell Drake," *Crossville (TN) Chronicle,* July 1, 2001; "Takes New Post," *MNJ,* Jan. 3, 1938; "Bromfields Move to Richland Farm by End of Month," *MNJ,* Mar. 16, 1939.

29. For more on Herschel Hecker, see "Herschel E. Hecker," *MS,* Mar. 14, 1991; "Two Will Serve Camp Ross Staff," *Chillicothe (OH) Gazette,* June 6, 1939; "Hecker Quits Soil Post," *MNJ,* Sept. 16, 1950; "Ohioan Is Named to Kentucky Job by U.S. Bureau," *Cincinnati (OH) Enquirer,* June 13, 1957. For more on demonstration projects and the CCC, see Hugh Bennett, *Soil Conservation,* 317–21; H. Wayne Pritchard, "Soil Conservation," in Clepper, *Origins of American Conservation,* 96; Maher, *Nature's New Deal,* 103–54; Lacy, *Soil Soldiers,* 152–54, 208; Sampson, *For Love of the Land,* 8–13; Hugh Bennett, "The Program Develops," *Hugh Bennett Lectures,* 22–23; Brinkley, *Rightful Heritage* 171, 202, 520–28; Helms, *Readings,* 7, 45–49; Hardin, *Politics of Agriculture,* 27. For specific information on the Rocky Fork camp, which was open from 1940–1942, see "12-Acre Site Offered for CCC Camp," *MNJ,* Jan. 10, 1940; "Up and Down the Street," *MNJ,* Sept. 12, 1940; "Give First Aid Instructions to CCC Members," *MNJ,* Sept. 27, 1940; "CCC Invites Public to Camp," *MNJ,* Apr. 1, 1941; "Open House," *MNJ,* May 7, 1941; "Up and Down the Street," *MNJ,* Jan. 26, 1943.

30. Scott, *Louis Bromfield,* 379–81; "Max Drake Remembers Bromfield," 161; R. R. Barker and D. T. Hover to Max Drake, May 8, 1940, box 92, folder 1572, LBC; Bromfield, *Pleasant Valley,* 103–5. For more on land use capability classifications and conservation plans, see Sampson, *Love of the Land,* 17–19; Ervin J. Utz, "The Coordinated Approach to Soil-Erosion Control," in USDA, *Soils and Men,* 666–78; Edward Graham, *Natural Principles,* 92–102; Headley, "Soil Conservation," 188–204.

31. Cooperative Agreement between Bromfield and SCS, box 92, folder 1572, LBC; Bromfield, *Pleasant Valley,* 104.

32. Bromfield, "A Piece of Land," *Land* 1, no. 3 (Review of Summer 1941): 206.

## 2. Crusading for Conservation

1. "Bromfield's Travels," *Land* 2, no. 2 (Review of Winter 1941–42): 42–43; "What They Said in Detroit," *Land* 3, no. 2 (Proceedings and Comment 1943–44): 221; "Proceedings," *Land* 6, no. 1 (Spring 1947): 248; Mary Bromfield, "The Writer I Live With," *Atlantic,* Aug. 1950, 77–79; Margaret Grabman, "The Fifth Horseman," *Land* 3, no. 2 (Proceedings and Comment 1943–44): 221.

2. "Proceedings," *Land* 2, no. 2 (Review of May, June, July 1942): 134.

3. Lord, *Behold Our Land,* 305. For biographical information on Russell Lord, see Eppig, "Russell Lord" and Lord and Lord, *Forever the Land.*

4. Jonathan Forman, "Earth and the Human Crop," *Land* 4, no. 4 (Review of Autumn 1945): 372; Hugh H. Bennett, "The Father of Friends of the Land," *Land* 7, no. 4 (Winter 1948–49): 484; Lord and Lord, *Forever the Land,* 39–42; Russell Lord, "Who Named These Friends?" *Land* 3, no. 3 (Review of Spring 1944): 322; Jonathan Forman and Ollie Fink, "Friends of the Land and Its Annual Conferences on Conservation, Nutrition, and Health—an Historical Sketch," in Forman and Fink, *Soil, Food, and Health,* 24; Morris L. Cooke, in "Proceedings," *Land* 1, no. 1 (Winter 1941): 14. For biographical information on Charles Holzer, see M. Whitcomb Hess, "Holzer of Gallipolis," *Land* 7, no. 2 (Summer 1948): 219–23; "Charles Holzer Biographical Sketch," box 11.3, FOTL; and Jonathan Forman, "An Appreciation of the Life and Works of Charles E. Holzer, Sr.," *LW* 2, no. 4 (Winter 1956): 3–4. For biographical information on Bryce Browning, see Chuck Martin, "Banker Bryce Browning Made His Mark in Conservancy," *ZTR,* Nov. 13, 1999, and "Bryce Browning Sr. Dies," *ZTR,* Mar. 13, 1984. For more on the history of the Muskingum Watershed Conservancy District, see "The Muskingum Story," *LW* 1, no. 1 (Spring 1955): 38–40; Bryce Browning, "The Muskingum Watershed Conservancy District," in Forman and Fink, *Water and Man,* 197–213; Bromfield, *Out of the Earth,* 173–85; and Bromfield, *From My Experience,* 220–34.

5. "Manifesto," *Land* 1, no. 1 (Winter 1941): 11–13; Lord and Lord, *Forever the Land,* 43–46.

6. Lord and Lord, *Forever the Land,* 52–66; Forman and Fink, "Friends of the Land," 24; Bromfield, *Pleasant Valley,* 278.

7. "Water Conservation Problem to be Discussed at Friends of the Land Parley Next Week-End," *CD,* July 13, 1941; "Friends of Land to Inspect Malabar as Part of Ohio Tour," *MNJ,* July 9, 1941.

8. Lord and Lord, *Forever the Land,* 114; Louis Bromfield, "A Piece of Land," *Land* 1, no. 3 (Review of Summer 1941): 203–6.

9. Russell Lord, "Field Notes: The Visit to Ohio," *Land* 1, no. 3 (Review of Summer 1941): 227–28; Lord and Lord, *Forever the Land,* 111–12; Louis Brom-

field, "New Pioneer of the Land," *Reader's Digest,* May 1945, 61–64; Bromfield, *Pleasant Valley,* 277–78; Frank Crow, "The Land Needs Friends," *Land* 1, no. 3 (Review of Summer 1941): 234–35; Harmel, Bonta, and Richardson, "Experimental Watersheds"; Helms, *Hydrologic and Hydraulic Research.*

10. Funderburk, *Conservation Education,* 30; Hammerman, "Historical Analysis," 87–126; Great Plains Committee, *Future of the Great Plains,* 121–27; Kaikow, "Legal and Administration Status," 3–43. For biographical information on Fink, see Burns Harlan, "Of People and the Times," *ZTR,* Feb. 14, 1965 and "Ollie Fink Dies at 71," *ZTR,* Mar. 1, 1970.

11. For more on the conservation education philosophy, see Studebaker, "Summary of Proceedings," Sept. 1937, box 103, FOTL; Bristow and Cook, *Education Program,* 4–13; "Educational News," 649; Fink, *Conservation for Tomorrow's America,* 3–9; Fink, "Developing the Program," 128; Funderburk, *Conservation Education,* 30–37; Kaikow, "Legal and Administration Status," 3–9; National Wildlife Federation, *Conference on Education in Conservation;* National Wildlife Federation, *Education in Conservation;* and Ward, Sears, and Ballam, *Foundations of Conservation Education.*

12. Ollie Fink to J. I. Falconer, Nov. 29, 1938, box 103, Ollie E. Fink, "Gone with the Wind and Water," pamphlet written for Zanesville Public Schools, 1938, box 3, Ollie Fink, "Conservation in Zanesville Schools," undated manuscript, box 3, Ollie Fink, *The Teacher Looks at Conservation* (Zanesville: Zanesville Public Schools, March 1939), box 3, Ollie Fink to Edmund Secrest, Aug. 14, 1939, box 24, Ollie E. Fink, "Developing the New Program of Conservation Education in Ohio," reprinted from *Ohio Schools,* January 1940, box 79, Ollie E. Fink, *The Teacher Looks at Conservation* (Columbus: Ohio Division of Conservation and Natural Resources, 1940), box 79, all in FOTL; O. E. Fink, "The Zanesville Plan Uses Facilities of the Muskingum Conservancy District," in National Wildlife Federation, *Education in Conservation,* 42–45; Kaikow, "Legal and Administration Status," 318–19, 326–29; Carl Johnson, "Conservation Education in Ohio," 39–40; Deron Mikal, "Of People and the Times," *ZTR,* June 16, 1974; Fink, *Conservation for Tomorrow's America.*

13. Eckelberry, "Teacher Education," 39–41; Johnson, "Conservation Education in Ohio," 41–44; Brink, "Way of Teaching"; Fink, *Conservation for Tomorrow's America,* 92, 141; Charles, "Tar Hollow," 104; Brink, "Clinical Tests," 78; Kaikow, "Legal and Administration Status," 319–21; O. E. Fink, "The Living Teacher," *Land* 3, no. 4 (Review of Summer 1944): 438–39; "Conservation Laboratory for Teachers," promotional poster, and Mildred Schmidt, "Possible Outcomes from My Study of Conservation at Tar Hollow," undated manuscript, both in box 23, FOTL.

14. Bromfield, *Out of the Earth,* 258; Phillips, *This Land,* 75; "Manifesto," 11–12. For a fuller discussion of this hypothesis, see Bromfield, *Pleasant Valley,* 119–20 and Bromfield, *Malabar Farm,* 103–5, 323.

15. Jonathan Forman, "Earth and the Human Crop," *Land* 4, no. 4 (Review of Autumn 1945): 373; Russell Lord, "Personal Mention," *Land* 3, no. 3 (Review of Spring 1944): 330–31; Forman and Fink, "Friends of the Land," 21. For biographical information on Jonathan Forman, see Forman, "How I Came into Medicine"; Helen Marsh, "Jonathan Forman, 1887–1967," folder 1, Jonathan Forman Papers; Harris, "Salute to Jonathan Forman," 706–10; "Dr. Forman," *CD,* Oct. 11, 1974; "Funeral Set for Forman," *CD,* Oct. 11, 1974; "Dr. Forman Dies at 87," *MNJ,* Oct. 12, 1974.

16. Carpenter, "Nutritional Science: Part 2"; Carpenter, "Nutritional Science: Part 3"; Renner, "Conservative Nutrition," 8, 139–42; Price, *Nutrition and Physical Degeneration;* McCollum, "Diet and Dental Caries"; McCarrison, *Nutrition and Health;* McCarrison, *Sir Robert McCarrison.*

17. Beeson and Matrone, *Soil Factor,* 78–113, 127–28; Huffman and Duncan, "Nutritional Deficiencies," 475; USDA, *Nutritional Quality of Plants,* 5–10; USDA, *Nutritive Value of Foods,* 5–7; Albrecht, *Soil Fertility and Animal Health.*

18. For details on Albrecht's experiments and contemporary criticism, see Albrecht and Smith, "Biological Assays"; Smith and Albrecht, "Feed Efficiency"; McLean, Smith, and Albrecht, "Biological Assays"; Beeson, "Effect of Mineral Supply," 442–44; and Iversen, "Critical Evaluation," 150. For more on the USDA's Plant, Soil, and Nutrition Laboratory (now the Robert W. Holley Center for Agriculture and Health), see USDA, *Nutritive Value of Foods,* ii, 2–3; Sandy Hays, "Plant/Soil/Nutrition Connection," 16–17; Beeson and Matrone, *Soil Factor,* 7–8; and Beeson, *Mineral Composition,* 2, 51–52, 56–57.

19. Forman and Fink, "Friends of the Land," 19–29; Paul Bestor, "Friends of the Land To-day," in Forman and Fink, *Soil, Food and Health,* 31–33; "Soil, a Dynamic Community," *Land News* 10, no. 3 (Autumn 1950): 16–23, 33; "Health: From the Ground Up," *Land News* 12, no. 3 (1952): 7–10, 33. The two books of conference proceedings are Forman and Fink, *Soil, Food and Health* and Forman and Fink, *Water and Man.*

20. Russell Lord, "A Day Afield," *Land* 2, no. 2 (May–July 1942): 157–58; Lord and Lord, *Forever the Land,* 183–86; Russell Lord, "Proceedings," and James Pope, "To Save Land, Control Water," both in *Land* 2, no. 3 (Review of Fall and Winter 1942–43): 177–80, 221–22.

21. For more on the TVA and its agricultural demonstration projects, see Judson King, *Conservation Fight;* Callahan, *TVA,* 152; Kyle, *Building of TVA;* Phillips, *This Land,* 83–107; Lilienthal, *TVA,* 79–92; Lewis Nelson, *U.S. Fertilizer Industry,* 208–13, 253; Wengert, *Valley of Tomorrow;* Lord, *Care of the Earth,* 316; Huxley, *TVA,* 28–49; Chandler, *Myth of TVA,* 101–3; Russell Lord, "Man of 1946," *Land* 5, no. 2 (Summer 1946): 181; and Lord, "Proceedings," 229–32.

22. Wengert, *Valley of Tomorrow,* 136; Chandler, *Myth of TVA,* 106–11; Bonita Irwin, "Wheat: The Community Hearts, Minds Don't Forget," *OakRidger,* Oct. 2, 2012, http://www.oakridger.com/article/20121002/NEWS/121009983; John Huotari, "Did You Know? Wheat Was Famous for Its Peach Orchards," *Oak Ridge Today,* Oct. 24, 2017, http://oakridgetoday.com/tag/dyllis-peach-orchard/; "A Brief History of Wheat," K-25 Virtual Museum, http://k-25virtualmuseum.org/happy-valley/wheat.html; Robinson, *Oak Ridge Story,* 36–37; Patricia A. Hope, "The Wheat Community," in Overholt, *Our Voices,* 14–19; Lord, "Proceedings," 233–35.

23. Lord, "Proceedings," 223–34; Robinson, *Oak Ridge Story,* 27; Atomic Heritage Foundation, "Civilian Displacement: Oak Ridge, TN," https://www.atomicheritage.org/history/civilian-displacement-oak-ridge-tn; Johnson and Jackson, *City behind a Fence,* 39–45; Hales, *Atomic Spaces,* 11–12, 50–56; John Rice Irwin, "New Places, Strange People," Jane Barnes Alderfer, "Oak Ridge in Contrast," Horace V. Wells Jr., "New Neighbors Come to Anderson County," and Thomas W. Thompson, "The Lost World of Black Oak Ridge," all in Overholt, *Our Voices,* 21, 100–106, 207–13, 246–53.

24. Lord and Lord, *Forever the Land,* 204; Hales, *Atomic Spaces,* 12; Callahan, *TVA,* 131; Johnson and Jackson, *City behind a Fence,* 6.

25. Lord and Lord, *Forever the Land,* 67–72.

26. Lord and Lord, *Forever the Land,* 73–74, 107–9; Russell Lord, "Notes Near Home: Some Further Pages from a Personal Yearbook," *Land* 7, no. 1 (Spring 1948): 85–86; "Notes and Letters," *Land* 2, no. 2 (Spring 1941): i, 124, 130.

27. Louis Bromfield to Charles Holzer, Feb. 4, 1942, Louis Bromfield to Charles Holzer, Aug. 3, 1942, Russell Lord to Bryce Browning, Aug. 5, 1942, Bryce Browning to Russell Lord, Aug. 10, 1942, and Russell Lord to Friends of the Land, May 25, 1942, all in box 15, FOTL; Lord and Lord, *Forever the Land,* 178; Russell Lord, "A Note on Business Matters," *Land* 2, no. 1 (Review of Winter 1941–42): 73–74.

28. Louis Bromfield, "Friends of the Land," *Reader's Digest,* Jan. 1944, 61–64; "Board of Directors," *Land Letter* 8 (Dec. 1948): 12; "Personal Mention," *Land* 3, no. 3 (Review of Spring 1944): 330–31; "An Independent Headquarters Established," *Land Letters* 4, no. 2 (1944): 1, 3; Forman and Fink, "Friends of the Land," 29; "Friends of Land U.S. Headquarters Open Here May 1," *CD,* Apr. 13, 1944; "Ollie Fink to New Post," *ZTR,* Apr. 12, 1944; "Proceedings: The First Five Years," *Land* 5, no. 1 (Review of Winter 1945–46): 101.

29. Russell Lord, "Where Bureaucracy and Democracy Meet," *Land* 4, no. 4 (Review of Autumn 1945): 480; Bromfield, "Friends of the Land."

### 3. A New Kind of Pioneer

1. "1,000 to Have Fun Tonight at Malabar War Relief Party," *MNJ,* June 7, 1941; "Mansfield News in Brief," *MNJ,* Aug. 30, 1941; Phil Dietrich, "Fishing Lines!" *ABJ,* Aug. 20, 1942; "Up and Down the Street," *MNJ,* Aug. 26, 1943; "Malabar Farm Is Scene of Concert," *MNJ,* June 20, 1941.

2. Bromfield, *Pleasant Valley,* 172–73; Little, *Green Fields Forever,* 28–30; Lal, Reicosky, and Hanson, "Evolution of the Plow," 3; Louis Bromfield. "The Evangelist of Plowman's Folly," *Reader's Digest,* Dec. 1943, 35–39.

3. Faulkner, *Soil Development,* 30–39, 58–63; Faulkner, *Second Look,* 12, 19, 29–32, back cover flap; Savoie Lottinville, foreword to Faulkner, *Plowman's Folly,* vii, 40–61; Bromfield, *Pleasant Valley,* 174; Beeman, "Trash Farmer," 91–93.

4. Bromfield, *Pleasant Valley,* 172–76; Faulkner, *Plowman's Folly,* 15, 77–80, 127–51; Little, *Green Fields Forever,* 32; Bromfield, *Malabar Farm,* 38.

5. William A. Albrecht, "The Indictment Will Not Stand," Hugh H. Bennett, "The Abolition of the Moldboard," and F. L. Duley, "No Control or Check . . . Too Many Assumptions," all in *Land* 3, no. 1 (Review of Summer 1943): 72, 66–67, 74–75; Faulkner, *Second Look,* 3–9.

6. Harold Martin, "Man with a Bull-Tongue Scooter," *Land* 3, no. 3 (Review of Spring 1944): 281–86; Christopher M. Gallup, "Another Plow-Free Farmer," *Land* 3, no. 2 (Proceedings and Comment, 1943–44): 160–61; "Proceedings," *Land* 3, no. 4 (Review of Summer 1944): 410–17; Allen and Fenster, "Stubble-Mulch Equipment," 11; Dillard, "Chisel Plow Dedicated"; American Society of Agricultural and Biological Engineers, "Graham-Hoeme Chisel Plow," http://www.asabe.org/About-Us/About-ASABE/History/ASABE-Historic-Landmarks/Graham-Hoeme-Chisel-Plow-2000.

7. Bromfield, *Pleasant Valley,* 180–204; Seaman Rotary Tiller brochure, box 136, folder 2220, LBC; Bromfield, *Out of the Earth,* 159–60.

8. Bromfield, *Pleasant Valley,* 204–9; Bromfield, *Malabar Farm,* 136–43, 256; Bromfield, *Out of the Earth,* 89, 121, 146–60.

9. Gardner, *American Agriculture,* 11–20; Holbrook, *Machines of Plenty,* 151, 166–67; Conkin, *Revolution Down on the Farm,* 16–18; Hurt, *Problems of Plenty,* 49; Scott, *Louis Bromfield,* 331–33.

10. Fraser, *Tractor Pioneer,* 17–23, 31–41, 67–80; Lyons, "Harnessing Power," 31; Neufeld, *Global Corporation,* 96–99,

11. Fraser, *Tractor Pioneer,* 90–118, 144–48; Gibbard, *Ford Tractor Story,* 153–55; Neufeld, *Global Corporation,* 100–107; Williams, *Massey-Ferguson Tractors,* 63–70; Williams, *Ford and Fordson Tractors,* 90–94.

12. Fraser, *Tractor Pioneer,* 148–55, 216; Bromfield, "Man with an Idea," circa 1944, box 128, folder 2108; Harry Ferguson to Louis Bromfield, May 29, 1945, box 129, folder 2111; Harry Ferguson, outline of "The Plan," circa 1943, box 129, folder 2111, all in LBC.

13. Louis Bromfield, "Can the Farm Catch Up with the Machine Age?" *Reader's Digest,* Oct. 1944, 77–79; Harry Ferguson to Louis Bromfield, Apr. 6, 1944, box 128, folder 2108, J. L. McCapprey to Reader's Digest, Oct. 24, 1944, box 129, folder 2109, Louis Bromfield, "What Agriculture Means to Detroit," address to the Economic Club of Detroit, Apr. 2, 1945, box 59, folder 1399, and Harry Ferguson to Louis Bromfield, Apr. 20, 1945, box 129, folder 2111, all in LBC; Bromfield, *Wealth of the Soil,* 27–29.

14. Gibbard, *Ford Tractor Story,* 157–61; Williams, *Massey-Ferguson Tractors,* 48, 75–80; Fraser, *Tractor Pioneer,* 173–240; Neufeld, *Global Corporation,* 64–68, 108–10, 128–47; Williams, *Ford and Fordson Tractors,* 103–5; Louis Bromfield to Roger Kyes, Sept. 23, 1944, box 129, folder 2111, Herman G. Klemm to Louis Bromfield, Jan. 13, 1954, box 129, folder 2110, and H. H. Bloom to Louis Bromfield, Apr. 14, 1954, box 129, folder 2110, all in LBC.

15. Bromfield, *Pleasant Valley,* 161–63; Bromfield, *Malabar Farm,* 119–21, 150; Ohio Auditor of State, audit of ODNR's operation of Malabar Farm from July 1, 1972 to Nov. 2, 1975, approved June 23, 1977, box 13, MF.

16. Bromfield, *Pleasant Valley,* 161–64; Bromfield, *Malabar Farm,* 112–24, 263; Bromfield, *Out of the Earth,* 68–74.

17. Bromfield, *Pleasant Valley,* 52, 319.

18. Hendrickson, *Food "Crisis,"* 2, 260; Fite, *American Agriculture,* 21; Bennett and Pryor, *This Land We Defend,* 103.

19. Tolley et al., "Agriculture in the Transition," 390; Elmer T. Peterson, "We Must Rebuild America," *Land* 3, no. 3 (Review of Spring 1944): 314–16; Chester C. Davis, "Building a Permanent Agriculture," *Land* 3, no. 4 (Review of Summer 1944): 449–51; Henry Bailey Stevens, "Fertility Down the Drain," *Land* 4, no. 1 (Winter 1945): 25–26; Louis Bromfield, "Saving This Land of Ours," address given at Friends of the Land annual meeting, Oct. 18, 1945, box 20, FOTL; Bromfield, *Pleasant Valley,* 319.

20. Maxine Jackson Kyle to Ollie E. Fink, Apr. 12, 1947, box 17, FOTL; Peggy Maupin, "'Pleasant Valley Is a Real Farm," *Denison (TX) Herald,* Aug. 13, 1946.

21. Geld, *Heritage,* 123–32; Margaret Suhr Reed, "Sunday Is Open House Day at Louis Bromfield's Farm," *CPD,* July 6, 1945; Louis Bromfield transcription no. 116, "Come and See," box 43, folder 606, Elden R. Groves, "Farmer Bromfield: A Visit with the Owner of Malabar Farm," *Farm and Dairy (Salem, OH)*, Dec. 5, 1946, box 131, folder 2135, and "Bromfield—and Malabar Farm," undated manuscript, box 132, folder 2137, all in LBC; William Shaw, "Louis Bromfield, Squire of Malabar Farm," *Scroll of Phi Delta Theta* 71, no. 2 (Nov. 1946): 99–106,

box 32, and Paul B. Sears, "What about Malabar?" undated manuscript, box 13, both in FOTL.

22. Bill Zipf, "Bromfield Knows His Farming," *CD,* June 29, 1947; Louis Bromfield, "Malabar Journal," undated manuscript, box 124, folder 2059, LBC.

23. Kyle to Fink, Apr. 12, 1947; Shaw, "Louis Bromfield"; Groves, "Farmer Bromfield."

24. Shaw, "Louis Bromfield," 99; Bromfield, *Pleasant Valley,* 214–40; Bromfield, *Malabar Farm,* 180–92; "Author Bromfield Dives under Water to Save Pet Dogs," *Logan (OH) Daily News,* July 9, 1946; Inez Robb, "If You Don't Like Dogs, Then—Don't Visit the Malabar Farm," *CD,* Jan. 11, 1949; Inez Robb, "Bromfield's Pet Boxers Get Plenty of Attention," *MS,* Sept. 15, 1950.

25. Groves, "Farmer Bromfield"; Bromfield, *Pleasant Valley,* 241–46; Bromfield, *Malabar Farm,* 166–74; Geld, *Heritage,* 100; Mary Bromfield, "Pleasant Valley . . . No Vacancies," *Vogue,* Dec. 15, 1946, 101, 118, 120, 125.

26. Geld, *Heritage,* 5–7, 131, 154; Bromfield, *Malabar Farm,* 45–46; George Hawkins, "Now Is the Time for All Good Men to Come to the Aid of Their Country," undated manuscript, box 16, FOTL; "Communications," *Land* 7, no. 1 (Spring 1948): 1.

27. Maupin, "'Pleasant Valley' Is a Real Farm"; Groves, "Farmer Bromfield."

28. Bromfield, *Malabar Farm,* 46–51.

#### 4. The Golden Age of Grass

1. Bromfield, *Malabar Farm,* 43–65, 112–65; Bromfield, *Out of the Earth,* 194–202.

2. Bromfield, *Malabar Farm,* 135–48; Robert F. Barnes and C. Jerry Nelson, "Forages and Grasslands in a Changing World," in Barnes et al., *Forages,* 3–23; Dale and Brown, *Grass Crops,* 1–2; Riedman, *Grass,* 17; Serviss and Ahlgren, *Grassland Farming,* 1–6; Staten, *Grasses and Grassland Farming,* xiii; Clinton P. Anderson, foreword to USDA, *Grass,* v; Jeremy W. Singer, Alan J. Franzluebbers, and Douglas L. Karlen, "Grass-Based Farming Systems: Soil Conservation and Environmental Quality," in Wedin and Fales, *Grassland,* 125–28.

3. Everett E. Edwards, "The Settlement of Grasslands," in USDA, *Grass,* 19–25; Vivien Gore Allen and John Herschel Fike, "H. A. Wallace, J. J. Ingalls, O. S. Aamodt, and Other Voices of the Nineteenth and Twentieth Centuries," in Wedin and Fales, *Grassland,* 16; Bromfield, *Malabar Farm,* 255. Bromfield contributed to Staten, *Grasses and Grassland Farming;* Rehm, *Twelve Cows;* Cope, *Front Porch Farmer;* and Allred and Dykes, *Flat Top Ranch.*

4. Cope, *Front Porch Farmer,* xxii–xxvii; "Georgia Revisited," *Land* 5, no. 1 (Review of Winter 1945–46): 121–25; Alderman, "Channing Cope"; Forseth and Innis, "Kudzu."

5. Quoted in Cope, *Front Porch Farmer,* 39–41.

6. Bromfield, *Malabar Farm,* 372–73; Louis Bromfield transcription no. 14, "Multiflora Rose," box 42, folder 603, LBC; Anderson and Edminster, *Multiflora Rose;* Ohio Invasive Plants Council, *Invasive Plants of Ohio Fact Sheet 8: Multiflora Rose,* www.oipc.info/uploads/5/8/6/5/58652481/8factsheetmultiflorarose.pdf.

7. Bromfield, *Out of the Earth,* 203; Howard Lytle, "The F. F. A. and Farmers' Class Trip to Malabar Farms," undated manuscript, box 121, folder 2028, and Louis Bromfield transcription no. 44, "Understanding Livestock," box 42,

folder 603, both in LBC; USDA Economic Research Service, "Dairy Products," Food Availability Per Capita, https://www.ers.usda.gov/data-products/food-availability-per-capita-data-system/; Baker and Falconer, *Costs of Producing Milk,* 6, 12; Shaudys, "Critical Analysis," 57.

8. For more on milk production before sanitary reform, see DuPuis, *Nature's Perfect Food;* Pirtle, *Dairy Industry;* Rawlinson, *Make Mine Milk;* Schlebecker, *American Dairying;* Smith-Howard, *Pure and Modern Milk;* and Valenze, *Milk.*

9. Weimar and Blayney, *Landmarks,* 3, 7; Pirtle, *Dairy Industry,* 130; Rawlinson, *Make Mine Milk,* 22–24; Heckman, "Securing Fresh Food," 475–76; DuPuis, *Nature's Perfect Food,* 74–78; Jones, *One Hundred Years,* 18; Herrington, *Milk and Milk Processing,* 166.

10. Eckles and Anthony, *Dairy Cattle,* 509–17, 577; McMurry, "Impact of Sanitation Reform," 26, 40; Louis Bromfield, "Maximum Dairy Profits and Health at Minimum Costs of Labor, Feed and Fertilizer," undated manuscript circa 1948, box 30, folder 455, LBC; Bromfield, *Out of the Earth,* 106–7; Bromfield, *Malabar Farm,* 130.

11. Petersen and Field, *Dairy Farming,* 465–71; Bromfield, "Maximum Dairy Profits"; Lytle, "Class Trip to Malabar"; Louis Bromfield transcription no. 37, "What Cows Like," box 42, folder 603, LBC.

12. Lytle, "Class Trip to Malabar"; McBride, *Ohio Farmer,* 13–14; Herrington, *Milk and Milk Processing,* 170–81, 219; McMurry, "Impact of Sanitation Reform," 27–34; Petersen and Field, *Dairy Farming,* 382; Milk receipts, box 132, folder 2143, 1951 dairy receipts, box 133, folder 2157, and Louis Bromfield, "The New Agriculture: A Fabulous World," address given to Little Sioux chapter of Friends of the Land, Mar. 16, 1949, box 34, folder 502, all in LBC; Mitchell and Baumer, *Interim Report,* 7.

13. Bromfield, *Malabar Farm,* 52–56, 127, 148; Bromfield, *Out of the Earth,* 127, 206–7; Louis Bromfield transcription no. 87, "Pasture and Chopper Feeding," box 42, folder 604, LBC; Pratt et al., *Soilage and Silage.*

14. Bromfield, *Malabar Farm,* 52; Morison, *Newer Hay Harvesting,* 4; Louis Bromfield, "Cashing in on Grass Farming," undated manuscript, box 8, folder 110, LBC.

15. Bromfield, *Out of the Earth,* 200–224.

16. Hayden et al., *Hay-Crop Silage,* 4; Pratt, Washburn, and Rogers, *Comparative Palatabilities,* 3; Louis Bromfield, "Grass Silage," address given at the Convention of the National Association of Silo Manufacturers, Nov. 29, 1949, box 19, folder 293, LBC; Bromfield, *Malabar Farm,* 124–28.

17. Bromfield, "Cashing in on Grass Farming"; Bromfield, *From My Experience,* 55–57; J. B. Shepherd et al., "Ensiling Hay and Pasture Crops," in USDA, *Grass,* 187–88; Louis Bromfield transcription no. 56, "Trench Silage Program," box 42, folder 604, LBC.

18. Shepherd et al., "Ensiling Hay," 187–88; Shaudys, Sitterley, and Studebaker, *Costs of Storing,* 5, 13, 18; Bromfield, "Cashing in on Grass Farming"; Bromfield, "Trench Silage Program."

19. Bromfield, *Out of the Earth,* 62, 200–203, 226–27, 293; Bromfield, "New Agriculture."

20. Backer, "World War II"; Weber, "American Way of Farming," 108; Hurt, *Problems of Plenty,* 98; Cochrane, *American Agriculture,* 124; Fite, *American Agriculture,* 20; Black, *Food Enough;* Hendrickson, *Food "Crisis."*

21. Schlebecker, *History of American Dairying,* 29, 41–44; Smith-Howard, *Pure and Modern Milk,* 55–61; Valenze, *Milk,* 235–52; USDA Economic Research Service, "Dairy Products," *Food Availability (Per Capita) Data System,* https://www.ers.usda.gov/data-products/food-availability-per-capita-data-system/; Louis Bromfield, "A Dairy Farmer Answers Back," undated manuscript circa 1948, box 11, folder 136, and Louis Bromfield, "A Restriction of Oleo Profit Potential-Package Deception Is Asked by Butter Interests," manuscript for *Voice from the Country* syndicated newspaper column, Jan. 25, 1949, box 92, folder 1563, both in LBC.

22. Staples, *Birth of Development,* 69–94; Norris Dodd, "Food and Agriculture Organization," 81–84; Shaw, *World Food Security,* 3–31; Lindberg, "Food Supply," 183; Boyd-Orr, *Food and the People,* 33–37; Ruxin, "Hunger, Science, and Politics," 53–58; Boyd-Orr, *As I Recall,* 171–96.

23. Arthur P. Chew, "The Catch in Industrialism," *Land* 5, no. 2 (Summer 1946): 266–72; Arthur P. Chew, "The Catch in Industrialism II: Food and Empire," *Land* 6, no. 1 (Spring 1947): 57–62; Malthus, *Principle of Population,* 14–19. For more on the history of neo-Malthusian ideas, see Connelly, *Fatal Misconception;* Hoff, *State and the Stork;* Linnér, *Return of Malthus;* and Robertson, *Malthusian Moment.*

24. Burch and Pendall, *Population Roads,* 2–7, 29, 39, 50–52; Perkins, *Geopolitics and the Green Revolution,* 119; USDA, *Agricultural Land Requirements.*

25. Guy Irving Burch, "More People, Less Food," *Land* 5, no. 3 (Autumn 1946): 355–60; L. G. Ligutti, "Against Race Suicide," *Land* 7, no. 1 (Spring 1948): 54–56; Russell Lord, "The War at Our Feet," *Land* 7, no. 3 (Autumn 1948): 433–36; Russell Lord, "The Conservation Crisis—Or Isn't It?" *Land* 7, no. 4 (Winter 1948–49): 524–32; William J. Gibbons, "Conserve or Starve," *America* 79, no. 23 (Sept. 11, 1948): 491–92; Osborn, *Plundered Planet,* 36–40, 193–201; Robertson, *Malthusian Moment,* 42–43, 170; Vogt, *Road to Survival,* 193–94, 265–88.

26. Ollie Fink, "Democracy and Human Freedom are Products of Fertile Soil," 1952, box 79, and Edward Danrich, "Here Comes Ollie," reprint from *Challenge,* Mar. 1952, box 3, both in FOTL.

27. Louis Bromfield, "Can Our Earth Feed Its People?" *Rotarian* 72, no. 2 (Feb. 1948): 10–13; Bromfield, *Malabar Farm,* 211–13, 217–23; Louis Bromfield, "We Don't Have to Starve," *Atlantic,* July 1949, 57, LBC Box 58, Folder 1396; Bromfield, *Out of the Earth,* 315–39.

28. Merrill Bennett, "Population and Food Supply"; Hugh Bennett, "Adjustment of Agriculture," 174–79; Salter, "World Soil"; Kellogg, "Food Production Potentialities"; Tolley, "Farmers in a Hungry World"; Belasco, "Algae Burgers"; Olmstead and Rhode, *Creating Abundance.*

29. Margaret Mattox, "Foreign Students Frolic at Malabar," *MNJ,* June 22, 1949; Peggy Mattox, "Tour Malabar, but Talk about Politics," *MNJ,* June 30, 1953; "County Beekeepers Visit Malabar Farm," *MS,* June 20, 1949; "Garden Club Tours Malabar Farm," *MNJ,* July 11, 1949; "Veterans Visit Malabar Farm," *Logan (OH) Daily News,* Aug. 20, 1949; Howard Lytle, "The F. F. A. and Farmers' Class Trip to Malabar Farm," undated manuscript, box 121, folder 2028, LBC.

30. "Large Crowd Hears Novelist Discuss Conservation of Soil," *ZTR,* Sept. 28, 1949.

## 5. Conservation at a Crossroads

1. Russell Lord, "Seeing the Country" and "Proceedings," *Land* 6, no. 4 (Winter 1947–48): 416–22, 533–34, 537–49; Tipton, "Texas Witnesses Soil Revival Meetings," *Acco Press* 35, no. 12 (Dec. 1947): 1–7, box 139, folder 2255, LBC; "The 7th Annual Meeting a Great Success," *Land Letter* 8 (Jan. 1948): 3, 10–11.

2. Lord, "Seeing the Country," 421–26, 531–37; Tipton, "Soil Revival Meetings," 5–7.

3. "Board of Directors," *Land Letter* 8 (Dec. 1948): 12; "Both Sides," *Land Letter* 8 (June 1948): 4; "Editor's Mail Box," *Land Letter* 9 (Nov. 1949): 4; front matter, *Land News* 10, no. 1 (Mar. 1950): 3.

4. E. J. Condon, "Report to the Board of Directors Tenth Annual Meeting Friends of the Land," Sept. 28, 1950, box 79, FOTL.

5. Lord and Lord, *Forever the Land,* 330–32; "Seeing the Country," *Land* 7, no. 3 (Autumn 1948): 317–24; "Seeing the Country," *Land* 8, no. 2 (Summer 1949): 273–78; "The Oklahoma Re-Run," *Land Letter* 8 (Dec. 1948): 3; Morgan, *Governing Soil Conservation,* 135.

6. Brink, *Big Hugh,* 125–29; "Seeing the Country," *Land* 8, no. 2 (Summer 1949): 271–78; "Annual Dinner: Expert Fixes U.S. Deadline on Soil Saving," *Land Letter* 8 (Dec. 1948): 4.

7. Morgan, *Governing Soil Conservation,* 120–43; "The Society's Annual Meeting," *Land Letter* 8 (Dec. 1948): 5; "Seeing the Country," *Land* 7, no. 4 (Winter 1948–49): 473–78.

8. "The Action Program of Friends of the Land," promotional brochure circa 1951, box 138, folder 2251, Louis Bromfield to Ed Condon, January 2, 1952, box 144, folder 2308, and Louis Bromfield, "Farming Is the Greatest and Most Difficult of Professions," undated manuscript, box 15, folder 221, all in LBC.

9. Hamilton, "Agribusiness"; Davis and Goldberg, *Concept of Agribusiness,* 2, 19–20, 80–81; Davis and Hinshaw, *Farmer in a Business Suit,* 228; Louis Bromfield, "The New Agriculture—A Fabulous World," address given to Little Sioux chapter of Friends of the Land, Mar. 16, 1949, box 34, folder 502, LBC.

10. Geld, *Heritage,* 152–62; D. K. Woodman to Louis Bromfield, Apr. 9, 1948, and Hugh H. Bennett to Louis Bromfield, Apr. 12, 1948, both in box 106, folder 1767, LBC; "Bromfield Aide Dies in Gotham," *CD,* Apr. 10, 1948; "Lonesome for 'Big House,'" *Mount Vernon (OH) News,* June 29, 1972.

11. "Famed Novelist Established Demonstration Farm Here," *Wichita Daily Times* (Wichita Falls, TX), Apr. 24, 1949; "Challenge Met, Bromfield Plans New Malabar Farm," *MNJ,* Feb. 20, 1949; Glenn Shelton, "Bromfield and Wichitans Reach Agreement on Malabar Farm," *Wichita Daily Times,* Feb. 20, 1949; Snow, "Great Dream," 381–85.

12. "Huge Quits at Malabar," *MNJ,* Dec. 15, 1949; "Malabar Farm Extended to South," *MNJ,* Dec. 27, 1949; Snow, "Great Dream," 390–402; Scott, *Louis Bromfield,* 492–95, 554–56; "Bromfield, Texans Lock Horns," *MNJ,* May 20, 1953; "Texas Farm Not Like Malabar, Bromfield Says," *CD,* Feb. 12, 1954; "Ohio Land Is Much Better," *CD,* Feb. 18, 1954.

13. Shaudys, "Critical Analysis," 57; "Malabar Farm Milk Production (Sales)," box 133, folder 2155, LBC; Louis Bromfield, "General Farms are Obsolete," *Successful Farming,* April 1950: 33, 98–101, box 19, MF; "No Longer Democrat, Bromfield Declares," *MNJ,* Sept. 30, 1950; "$4,588 Spent by Taft Unit," *MNJ,* Nov. 21, 1950; Inez Robb, "Bromfield Now Wired for Sound," *CD,* Sept. 21, 1950.

14. "Semiannual Meeting of Board of Directors of Friends of the Land," *Land News* 11, no. 2 (1951): 9, 21; "War between Ohio, Kentucky Averted at Malabar Farm," *Lancaster (OH) Eagle-Gazette,* July 11, 1951; George Constable, "Bromfield Jibes at Canadian Grasses," *MNJ,* July 11, 1951.

15. Ken Davis, "The Fabulous Lives of Louis Bromfield," *CD,* Apr. 23, 1950; Inez Robb, "Eating Time at Malabar," *MNJ,* Sept. 14, 1950.

16. "Hope Bromfield and Robert Stevens Jr. Say Vows," *MNJ,* Dec. 24, 1950; "Ellen Bromfield Takes Vows in New York City," *MNJ,* Jan. 7, 1951; Robert Schweitz, "Douglas Farm Added to Malabar's Acres," *MNJ,* May 3, 1951; Geld, *Heritage,* 167–70; Ellen Geld, "It's Fun to Live in Your Own House," *MNJ,* June 22, 1952; Geld, *Strangers in the Valley,* 17–19; Ellen M. Geld, "Let Winter Come," *MNJ,* Nov. 11, 1951; Ellen M. Geld, "Farm Manager Puts Experience to Work," *MNJ,* Dec. 2, 1951.

17. "Group to Visit Malabar Farm," *ABJ,* July 19, 1951; Ellen M. Geld, "The World Rings Bromfield Doorbell," *MNJ,* Dec. 9, 1951; Ellen Geld, "Ballet Troupe Relaxes at Malabar," *MNJ,* Oct. 26, 1952; Virginia Lee, "Receives Moose Meat for Good Turn," *MNJ,* Dec. 2, 1951; Ellen M. Geld, "'Hang the Government' to the Tune of Brahms," *MNJ,* Dec. 23, 1951.

18. Ellen M. Geld, "Oven Blast, Train Make Holiday Hectic," *MNJ,* Dec. 30, 1951; Inez Robb, "Fit for a King," *MS,* Jan. 7, 1952; Ellen Geld, "Malabar Observes Thanksgiving Day," *MNJ,* Dec. 7, 1952; Lou Whitmire, "Former Malabar Farm Cook Dies at Age 107," *MNJ,* Aug. 8, 2014; Scott, *Louis Bromfield,* 363.

19. Program for *Successful Farming* field day at Malabar Farm, Aug. 9, 1952, box 131, folder 2127, LBC; "10,000 Attend Field Days," *MNJ,* Aug. 10, 1952; Ellen Geld, "Field Day Malabar's Biggest Farm Show," *MNJ,* Aug. 17, 1952; Bromfield, *From My Experience,* 42; "Mobile Unit Offers First Aid, Food," *MNJ,* Oct. 6, 1952.

20. Geld, *Heritage,* 168–83; Margaret Mattox, "Malabar Products Sold in New York," *MNJ,* Aug. 24, 1951; Ellen Geld, "Plum, Peach Butter Flow at Malabar," *MNJ,* Aug. 31, 1952; Geld, *Strangers in the Valley,* 16–22.

21. Geld, *Heritage,* 179–83; "Heart Attack Ends Life of Mrs. Bromfield," *MNJ,* Sept. 15, 1952; Ellen Geld, "Time Draws Near for Trip to Brazil," *MNJ,* Feb. 15, 1953; Ellen Geld, "Bids Final Farewell to Malabar Farm," *MNJ,* Mar. 22, 1953. For more on Malabar-do-Brasil, see Geld, *Strangers in the Valley;* Bromfield, *From My Experience,* 87–133; and DeVault, *Return to Pleasant Valley,* 291–95.

22. Minutes from Friends of the Land board of directors meeting at Topeka, Kansas, Mar. 23, 1954, box 109, FOTL; Jonathan Forman, Report to Directors, May 8, 1953, box 139, folder 2263, LBC; Jonathan Forman, "Growth and Change," *Land* 11, no. 4 (1953): 351.

23. Hunter Baker to Jonathan Forman, Dec. 3, 1952, and "The Role of Dr. Forman in the Production of the Land," memorandum, Jan. 1953, both in box 78, FOTL; "Across the Land," *Land* 11, no. 4 (1953): 353; "Why Seek a Publisher?" *Land* 12, no. 4 (Review of Winter 1953–54): inside front cover to 388.

24. McConnell, "Conservation Movement," 463; Pete E. Cooley, "It Pays—But You Can't Buy It," *Land* 12, no. 1 (Spring 1953): 88–89; "Tax Money and Land," *LW* 4, no. 1 (Spring 1958): 23; Hone, "Analysis of Conservation Education," 1–3, 85–86; Kaikow, "Legal and Administration Status," iv, x, 3–4, 633–37; Palmer, "Conservation Education," 194. For more on the politics of soil conservation in the 1950s, see Morgan, *Governing Soil Conservation,* and Hardin, *Politics of Agriculture.*

25. Hurt, *Problems of Plenty,* 98–114; Fite, *American Agriculture,* 24; Weber, "Manufacturing the American Way of Farming," 108; Iowa State, *Problems and*

*Policies,* 11; Schlebecker, *Whereby We Thrive,* 278; Shaudys, "Critical Analysis," 57, 75; Malabar Farm dairy receipts and expenses, 1951, folder 2157 and Louis Bromfield, "Bulletin 1: January 1952," folder 2159, both in box 133, LBC.

26. Cochrane, *American Agriculture,* 388, 427–29.

27. Russell Lord, "The Conservation Tangle," and C. L. Swanson, "Some Changing Ideas of Soil Conservation," both in *Land* 12, no. 1 (Spring 1953): 93, 95–99.

28. Meine, *Aldo Leopold;* Flader, *Thinking Like a Mountain;* Edward Graham, *Natural Principles,* 8, 92–104, 231; Stuckey, "Paul Bigelow Sears"; 104; Sinnott, "Paul B. Sears"; Gerald Young, *Human Ecology;* Sears, "Human Ecology," 961; Gordon, "Traditional Ecology," 490–91; Frederick Smith, "Ecology and the Social Sciences," 763–64; Sears and Carter, "Ecology and the Social Sciences," 300.

29. Sears, *Ecology of Man,* 8; Paul B. Sears, "Human Ecology," *Land* 10, no. 1 (Spring 1951): 23–26; Shepard and McKinley, *Subversive Science,* 1; Robert Smith, *Ecology of Man;* Louis Bromfield, "The Power of Example," *Land* 11, no. 4 (1953): 379–86; Jonathan Forman, "The Nutrition Conference," *Land News* 11, no. 3 (1951): 9; Louis Bromfield to Chester Davis, Ed Condon, and Paul Sears, box 144, folder 2308, LBC.

30. "Malabar to Become Farm 'Foundation,'" *MNJ,* Mar. 15, 1953; Louis Bromfield, "Malabar Farm," promotional pamphlet, 1953, MF; Louis Bromfield to Jonathan Forman, box 13, FOTL.

## 6. Vegetables on the Middle Ground

1. Bromfield, *From My Experience,* 119–20, 199–215, 273, 282; Geld, *Heritage,* 105–6. The water flow rate is taken from a plaque on the stand.

2. "Sell Malabar Vegetables at Stand," *MNJ,* July 2, 1954; Bromfield, *From My Experience,* 215, 275–76; Louis Bromfield transcription no. 40, "Going into the Roadside Market Business," box 42, folder 603, and Louis Bromfield, "Cooking at Malabar," undated manuscript, box 11, folder 133, both in LBC.

3. Lawrence M. Hughes, "Bromfield Puts 'Malabar Farm' into the Gift-Food Business," *Sales Management,* Oct. 1, 1953, 32–39; Virginia Lee, "Malabar Jams, Jellies Go on Sale," *MNJ,* Nov. 5, 1953; "Malabar Farm Calendar Now on the Market," *CD,* Dec. 23, 1953; Scott, *Louis Bromfield,* 613–23; Louis Bromfield, "Malabar Farm," promotional pamphlet, 1953, MF.

4. Bromfield to Monsanto Chemical Company, July 13, 1953, retyped in undated manuscript, box 40, folder 608, SP; Bromfield, *From My Experience,* 37–38; Ashby, "Aluminum Legacy"; Doordan, "Promoting Aluminum"; Branyan, "From Monopoly to Oligopoly"; Reynolds Metals Company Annual Reports, 1955 and 1956, *America's Corporate Foundation,* available online at ProQuest Historical Annual Reports, https://about.proquest.com/en/products-services/pq_hist_annual_repts/; Lauber, "It Never Needs Painting"; "Institute to Spur Farm Aluminum Use," *New York Times,* Mar. 1, 1950; "Farm Experts Swap Dreams of Turbo-Jet Tractors, Cow Cafeterias, Crop Dryers," *Wall Street Journal,* June 20, 1953.

5. Louis Bromfield to William G. Reynolds, Oct. 27, 1953, W. G. Reynolds to Bromfield, Apr. 21, 1954, and Louis Bromfield to Richard G. Moser, June 10, 1954, all in box 136, folder 2212, LBC.

6. Terry, "Thermodynamics of Hay Driers," 3; Ramser, Andrew, and Kleis, *Better Hay;* Louis Bromfield transcription no. 89, "More on Haydrying," box

42, folder 604, and Louis Bromfield, edits to Reynolds sales document, box 130, folder 2125, both in LBC.

7. Louis Bromfield, Report on Malabar-Reynolds Hay Drying Pilot Plant, box 134, folder 2177, and E. E. Wood to James Bernhardt, Aug. 9, 1954, box 130, folder 2125, both in LBC; Bromfield, *From My Experience,* 57–64.

8. Louis Bromfield transcriptions nos. 108, 109, and 110, "Haydrying," and 118, "Haydrier," box 43, folder 606, E. E. Wood to Louis Bromfield, Nov. 1, 1954, and June 17, 1955, box 130, folder 2125, and "Now—Produce Top Quality, High Protein Hay with a Reynolds Aluminum Self-Feeding Hay Drying Barn," promotional flyer, box 134, folder 2177, all in LBC; Bromfield, *From My Experience,* 57–64.

9. Bromfield, *From My Experience,* 15–32, 271–72; Frances B. Murphey, "Bromfield's Interest Turns to Vegetables, Cheesemaking," *ABJ,* Aug. 1, 1954; Virgil A. Stanfield, "Malabar Farm Garden," *MNJ,* July 14, 1974.

10. Malabar Roadside Market Partnership Contract, box 130, folder 2116, Sales from Roadside Stand, box 137, folder 2230, both in LBC; Bromfield, *From My Experience,* 272–76.

11. Bromfield, *From My Experience,* 197; Bromfield, *Malabar Farm,* 235; statement of Louis Bromfield, in *Chemicals in Food,* pt. 1:290–95; McWilliams, *American Pests,* 5–25, 42–46, 82–116; Whorton, *Before Silent Spring,* 8, 20–41; Edmund Russell, *War and Nature,* 21–23, 66, 77; Clarke, "Uncle Sam Raises Bugs."

12. Zimmerman and Lavine, *DDT,* 1–2, 42–43, 119; Dunlap, *DDT,* 62–63; Edmund Russell, *War and Nature,* 111; Whorton, *Before Silent Spring,* 249.

13. Frederick Davis, *Banned,* 38–71, 97–100, 147; testimony of Morton S. Biskind, in *Chemicals in Food* (1950), 700–722; Biskind, "DDT Poisoning"; Biskind and Bieber, "DDT Poisoning"; testimonies of Francis Marion Pottenger Jr. and Granville Frank Knight, both in *Chemicals in Food,* pt. 2:931–32, 1049–1051.

14. Wiley, *Autobiography;* Oscar Anderson, *Health of a Nation;* Hilts, *Protecting America's Health,* 89–94; White, "Chemistry and Controversy," 309–16; Kallet and Schlink, *100,000,000 Guinea Pigs;* Lamb, *American Chamber of Horrors;* Whorton, *Before Silent Spring,* 251–52; Frederick Davis, *Banned,* 19–22; *Chemicals in Food* (1950), 1.

15. Testimonies of Paul A. Neal and Faith Fenton, both in *Chemicals in Food* (1950), 137, 743; testimony of J. T. Sanders, in *Chemicals in Food,* pt. 1:355; testimonies of John R. Magness, Fred L. Overley, and Robert L. Webster, all in *Chemicals in Food,* pt. 2:607, 640, 655–57.

16. Louis Bromfield, in *Chemicals in Food,* pt. 1:289–309.

17. Howard, *Agricultural Testament,* ix, 53–56; Howard and Wad, *Waste Products;* Heckman, "History of Organic Farming," 144; Conford, *Origins,* 53–58; Howard, *Farming and Gardening,* 15–19, 216–19; Barton, *Global History,* 49–76, 96–103; Franklin King, *Farmers of Forty Centuries;* Waksman, Tenney, and Diehm, "Preparation of Artificial Manures"; Hutchinson and Richards, "Artificial Farmyard Manure"; Brady and Weil, *Elements,* 16–19.

18. Howard, *Agricultural Testament,* 1–25, 105–15; 139–80; Howard, *Farming and Gardening,* 5, 29, 73–74.

19. Bromfield, *Pleasant Valley,* 159, 210–12; Bromfield, *Out of the Earth,* 26; Louis Bromfield, "Foundation for Life," *New York Times,* Jan. 19, 1947; Howard, *War in the Soil,* 23.

20. Manlay, Feller, and Swift, "Soil Organic Matter Concepts," 221–22; Korcak, "Early Roots," 263–66; Van der Ploeg, Böhm, and Kirkham, "History of Soil

Science," 1058–61; Liebig, *Chemistry,* 29–32, 158–59, 179, 261; Liebig, *Natural Laws,* 19–20, 177, 183, 369; Johnston and Poulton, "Long-Term Experiments"; E. John Russell, "Experiment Station"; Lewis Nelson, *U.S. Fertilizer Industry;* Smil, *Enriching the Earth.*

21. E. John Russell, "Agricultural Science"; Waksman, *Humus;* Yoder, "Fertilizer Efficiency"; Wilson, "Conservation of Our Soil Resources," 3; Deitmeyer, "Wilfred M. Schutz," 2–3; Schutz, "First Place Winner," 4.

22. Margaret Merrill, "Eco-Agriculture," 194; Treadwell, McKinney, and Creamer, "Philosophy to Science," 1010; Blum, "Composting," 174; Conford, *Origins,* 103; Howard and Wad, *Waste Products,* 20; Howard, *Agricultural Testament,* 37–38, 186; Howard, *Farming and Gardening,* 15, 75–85; Howard, *War in the Soil,* 8, 60.

23. Rodale, *Organic Front,* 61; Ray I. Throckmorton, "The Organic Farming Myth," *Country Gentleman* 121 (Sept. 1951): 21, 103, 105; Emil Truog, "Organics Only? Bunkum!" *Land* 5, no. 3 (Autumn 1946): 317–20; Bear, "Facts . . . And Fancies." For biographical information on Rodale, see Case, *Organic Profit;* Jackson, *J. I. Rodale;* Conford, *Origins,* 100; O'Sullivan, *American Organic,* 17–67, 113, 132, 224–25; and Rodale, *Organic Merry-Go-Round,* 49–55.

24. Bromfield, *Malabar Farm,* 157–58, 283–304; Bromfield, "Foundation for Life"; Bromfield, *Out of the Earth,* 81–82.

25. "Seeing the Country," *Land* 10, no. 2 (1951), 237; "Organics Only?" *Land* 5, no. 1 (Review of Winter 1945–46): 45–61; Alden Stahr, "Natural Farming at Stahrland," *Land* 10, no. 2 (Summer 1951): 181–85; Alden Stahr, "Not so Cockeyed," *Land* 11, no. 1 (Spring 1952): 42–46.

26. David Greenberg, "What Is This Thing Called Organic Farming?" *Land* 10, no. 3 (Autumn 1951): 243, 349; "Views and Visits," *Land* 10, no. 4 (Winter 1951–52): 353–55, 418–22; "Views and Visits," *Land* 11, no. 1 (Spring 1952): 99–101; Jonathan Forman, "Soil and Man," *Land* 11, no. 2 (Summer 1952): 177–80; "Communications," *Land* 11, no. 3 (Jan. 1953): 337–38; "Communications," *Land* 12, no. 1 (Spring 1953): 103–6.

27. Bill Zipf, "New Plant Nutrition Plan Yields Big in Tests Here," *CD,* June 21, 1953; "Bromfield to Indorse New Fertilizer Here," *CD,* June 21, 1953; Fertileze advertisement, *CD,* Sept. 13, 1953; Bromfield, *From My Experience,* 146–97.

28. Bromfield, *From My Experience,* 35, 154.

### 7. Saving Malabar

1. "Plans Dairy Herd Sale," *MNJ,* Nov. 14, 1954; "Public Sale," *MNJ,* Nov. 16, 1954; Bob Liston, "'Quiet' Bidders Buy Bromfield's Herd," *MNJ,* Nov. 18, 1954; Walter D. Hunnicutt to Louis Bromfield, Dec. 3, 1954, box 143, folder 2301, LBC; "Veteran Dairy Farmer Works Many Hours, but Shows Profit," *MNJ,* Jan. 16, 1955.

2. "Bromfield Suffers 'Acute Infection,'" *CD,* Dec. 8, 1954; Ellen Bromfield Geld, "Bromfield Roams Brazilian Farm, Shows Recovery from Recent Illness," *MNJ,* Feb. 13, 1955; Geld, *Heritage,* 184–85; Bromfield, *From My Experience,* 294–301.

3. Schweitzer, *Life and Thought,* 154–59; Bromfield, *From My Experience,* 301–6.

4. Ollie Fink to Albert Beehler, July 10, 1953, box 43, and "Friends of the Land Board of Directors Meeting," Sept. 29, 1953, box 17, both in FOTL; "Why Seek a Publisher?" *Land* 12, no. 4 (Review of Winter 1953–54): inside front

cover to page 388; Friends of the Land Board of Directors Meeting, Mar. 23, 1954, and Jonathan Forman, Report to the Board of Directors, July 16, 1954, both in box 109, FOTL.

5. Memorandum of Understanding between Russell & Kate Lord and Dr. Jonathan Forman, Jonathan Forman, "A Memorandum of Understanding," May 13, 1954, and Russell Lord, Personal Memorandum, Mar. 5, 1960, all in box 109, FOTL; Alfred H. Williams, "Growth: A Statement of Enlarging Purposes," *Land* 13, no. 1 (Review of Spring 1954): 1–3; Minutes of the Meeting of the Land Trustees," Jan. 25, 1955, and Louis Bromfield to Monroe Bush, Feb. 18, 1955, both in box 139, folder 2262, LBC.

6. Lord, Personal Memorandum; Helms, *Readings,* 32; Sampson, *For Love of the Land,* 56; Louis Bromfield, Ollie E. Fink, and Jonathan Forman to "Dear Friends," *LW* 1, no. 1 (Spring 1955): inside front cover; "Editor's Mail Box," *LW* 1, no. 2 (Summer 1955): 2.

7. "The Real Task of Friends of the Land," *LW* 1, no. 1 (Spring 1955): back cover.

8. "Bromfield Quits Wildlife Council," *New Philadelphia (OH) Daily Times,* May 6, 1955; "Bromfield to Restore Old Inn," *MNJ,* June 26, 1955.

9. Louis Bromfield transcription no. 53, "On Farm Parties," box 42, folder 604, LBC; Joan Brown, "Bromfield, Friends Enjoyed Cabin Visits," *MNJ,* Apr. 21, 1974; Geld, *Heritage,* 172–73; Scott, *Louis Bromfield,* 609–11.

10. Walker, "Midas of the Texas Range," *Saturday Evening Post,* July 27, 1957, 22, 61–64; Allred and Dykes, *Flat Top Ranch;* Frank Reeves, "Ranchmen Visit Flat Top Ranch," *Cattleman* 36, no. 6 (Nov. 1949): 104–8; Charles Pettit to Louis Bromfield, Nov. 7, 1951, and Bromfield to Pettit, June 8, 1954, both in box 13, FOTL; Bromfield to Pettit, Apr. 3, 1955, box 107, folder 1778, LBC.

11. "Bromfield Denies 'Romantic Link' with Doris Duke," *New Philadelphia (OH) Daily Times,* Jan. 24, 1956; Geld, *Heritage,* 194–95; Louis Bromfield death certificate, MF; "Bromfield Ill at Malabar," *MNJ,* Feb. 15, 1956; "Report Bromfield Condition is 'Satisfactory,'" *CD,* Mar. 11, 1956; Ken Woodman, "Louis Bromfield, World-Famed Mansfield Author, Dies at 59," *MNJ,* Mar. 19, 1956; Scott, *Louis Bromfield,* 630–34.

12. "Nation's Leaders Pay Tribute," *MNJ,* Mar. 19, 1956; "500 Attend Service for Bromfield," *CD,* Mar. 23, 1956; "600 Attend Last Tribute to Bromfield," *CPD,* Mar. 23, 1956; Charles Pettit to Ann Rimmer, Mar. 21, 1956, box 13, FOTL; Ollie E. Fink, "A Man Who Loved the Soil," *Zanesville Times Signal,* Mar. 25, 1956; Inez Robb, "Cherished Friend Pens Tribute to Bromfield," *MNJ,* Mar. 25, 1956.

13. Margaret Mattox, "The Story of Bromfield's Malabar," *MNJ,* Mar. 25, 1956; "Daughters Inherit Bromfield Estate," *CD,* Mar. 27, 1956; "Group Would Use Malabar Farm as Research Center," *Coshocton (OH) Democrat,* Apr. 18, 1956; "In Appreciation of the Life and Work of Louis Bromfield," *LW* 2, no. 2 (Summer 1956): 3–6; "An Invitation . . . for You to Visit Battelle," promotional pamphlet, Nov. 1951, "Battelle: What It Is; How It Operates," promotional pamphlet, untitled promotional pamphlet about Battelle's agricultural research, and "Malabar Farm Research and Educational Institute," undated manuscript, all in box 40, folder 608, SP.

14. "Suggested Objectives for a Research Program for 'Bromfield Memorial Foundation," undated manuscript, circa 1956, folder 608, and "Proposal," undated manuscript, folder 609, both in box 40, SP; "Plans for Malabar," *LW* 2, no. 4 (Winter 1956): 5.

15. "Suggested Objectives"; "Proposal"; "The Battelle Plan of Research and Malabar Farm," undated manuscript, box 13, FOTL.

16. "Plans for Malabar"; "Proposal"; "An Ecologic Institute and Center," undated manuscript, and Paul B. Sears to Ollie E. Fink, Aug. 30, 1956, both in box 39, folder 602, SP.

17. Ellen Bromfield Geld, "Bromfield's Daughter Hopes to See Malabar Become Ecological Center," *MNJ*, Nov. 4, 1956; Hope Stevens to Bill Solomon, Nov. 19, 1958, NF; Paul B. Sears to Clyde Williams, Aug. 20, 1956, folder 602, Forrest D. Murden to Paul Sears, Aug. 31, 1956, folder 602, Paul B. Sears to Clyde E. Williams, Mar. 14, 1957, folder 603, Clyde Williams to Paul B. Sears, Mar. 20, 1957, folder 603, and Paul B. Sears to Clyde Williams, Apr. 4, 1957, folder 603, all in box 39, SP.

18. Ollie Fink to Ed Condon, Sept. 7, 1956, box 39, folder 602, SP; Bromfield, *From My Experience,* 35–39.

19. Bear, Toth, and Prince, "Variation in Mineral Composition," 381–83; "'Hollow Food': Some Questions and Comments," *Land* 8, no. 1 (1949): 51–55; USDA, *Nutritive Value of Foods,* 9, 18–19; Maynard, "Effect of Fertilizers"; Brandt and Beeson, "Influence of Organic Fertilization," 449; Beeson and Matrone, *Soil Factor,* 32–50.

20. Michigan State University, *Nutrition of Plants;* "Editorial," *LW* 1, no. 2 (Summer 1955): 3; Sears, "Soil and Health," 270.

21. Catherine Rorick, "They've Worked to Keep Malabar Alive," *MNJ,* May 11, 1957; Margaret Mattox, "Famed Malabar Farm for Sale," *MNJ,* Jan. 6, 1957; Marguerite Miller, "'Friends' Offer to Buy Malabar," *MNJ,* Jan. 23, 1957; Ollie E. Fink to Paul B. Sears, Feb. 4, 1957, box 39, folder 603, SP; "Robinson Proposes State University at Malabar Farm," *MNJ,* Jan. 30, 1957; "Robinson Pushes Malabar Purchase," *MNJ,* Mar. 19, 1957; "Sale of Malabar to Friends of the Land Reported Near," *CD,* Mar. 15, 1957.

22. Peter T. Lannan Jr., "Experimental Woods: Louis Bromfield's Malabar Farm, Lucas, Ohio," manuscript, June 1946, box 24, MF; Ollie Diller, "Ecological Pattern of the Malabar Woods," *LW* 4, no. 3 (Autumn 1958): 13–15; "The Louis Bromfield's Malabar Farm Foundation Inc.," *LW* 4, no. 4 (Winter 1958): 2; Oliver D. Diller to Bill Solomon and Louis Bromfield, Jan. 26, 1956, box 137, folder 2233, LBC; Jonathan Forman to Herbert Cobey and Bill Solomon, Feb. 20, 1959, box 17, FOTL; "Malabar Farm Trees Being Cut," *MNJ,* Mar. 18, 1957; "Lumbermen at Malabar Farm," *MNJ,* Mar. 19, 1957.

23. "Executive Committee Meeting of Purchase of Malabar," Apr. 3, 1957, box 109, FOTL; "Malabar Is Still without a Buyer," *CPD,* Apr. 26, 1957; "You Can Help Malabar Farm to Continue," Friends of the Land fundraising flyer, box 139, folder 2256, LBC.

24. Faulk, Faulk, and Gray, *Imagination and Ability;* Noble Foundation, *Tracing Our Steps;* Noble Foundation to Louis Bromfield, July 26, 1949, box 104, folder 1747, LBC; Francis J. Wilson to Louis Bromfield, Aug. 17, 1949, box 11.2, FOTL; "Background Information—The Samuel Roberts Noble Foundation, Inc. Ardmore, Oklahoma," manuscript, May 10, 1957, box 39, folder 603, SP; Scott, *Louis Bromfield,* 494.

25. "Five National Annual Conferences of Friends of the Land," *LW* 2, no. 4 (Winter 1956): 29; Notes for Conference with Mr. Forbes, box 109, FOTL; Ollie E. Fink to C. C. Forbes, May 3, 1957, and Ollie Fink to Cecil Forbes, telegram, May 6, 1957, both in NF.

26. Samuel Roberts Noble Foundation to First National Bank of Mansfield, May 6, 1957, and telegram, James E. Thompson to Richard G. Moser, May 6, 1957, both in NF; "Malabar Sale at $140,000 Near," *MNJ,* May 9, 1957; "Malabar Has Been Saved!" *LW* 3, no. 2 (Summer 1957): 1.

27. "Friends of the Land Get Malabar," *MNJ,* May 11, 1957; Emerson L. Batdorff, "Sale 'Saves' Malabar," *CPD,* May 11, 1957; Inez Robb, "Inez Writes of Malabar Sale and Ellen's Visit," *MNJ,* May 15, 1957; "Check for $30,000 Opens Drive," *ZTR,* May 13, 1957.

28. Ollie Fink and William D. Murphy to George E. Beach, May 20, 1957, box 13, FOTL; "Timberland at Malabar Farm Saved by Friends," *Galion (OH) Inquirer,* May 25, 1957; "Malabar Trees to be Saved," *MNJ,* May 26, 1957; "Malabar Farm Trustees Halt Cutting of Timber," *LW* 3, no. 2 (Summer 1957): 6.

29. "Friends of the Land Get Malabar," *MNJ,* May 11, 1957; Ollie Fink to Marquis Childs, May 24, 1957, and Paul Sears, "What about Malabar?" undated manuscript, both in box 13, FOTL; Ollie E. Fink, memorandum about Malabar fundraising campaign, July 8, 1957, folder 605, Paul B. Sears to Walter Pretzer, June 18, 1957, folder 604, and Ollie Fink to Paul B. Sears, June 21, 1957, folder 604, all in box 40, SP.

## 8. Marginalization

1. Bernard M. Shaaley to Ollie Fink, June 12, 1957, box 40, folder 604, SP; "Ike Praises 'Friends of the Land' Operations at Malabar Farm," *MS,* July 30, 1957; Dwight D. Eisenhower to Ollie Fink, Aug. 7, 1947, box 139, folder 2256, LBC.

2. "Malabar Farms Tours to Resume Sunday, Aug. 4," *Galion (OH) Inquirer,* July 25, 1957; "Malabar Farm Tours Will Be Resumed," *Mt. Vernon (OH) News,* July 24, 1957; "200 Present for Tea, Tour of Malabar Farm 'Big House,'" *BTF,* Sept. 10, 1957; "Open House Draws 2,000 to Malabar," *MNJ,* Sept. 15, 1957.

3. Freeman Lincoln to Ollie Fink, June 12, 1957, De Witt Wallace to Ollie Fink, July 31, 1957, Harper's magazine to Ollie Fink, Aug. 14, 1957, and Cass Canfield to Ollie Fink, June 25, 1957, all in box 13, FOTL.

4. Minutes of the 18th Annual Meeting of the Board of Directors of Friends of the Land, Sept. 19, 1958, and Malabar Farm Project Income and Expense Statements, Year Ended Aug. 31, 1958, both in NF; Jonathan Forman to Ollie Fink, Jan. 15, 1959, box 109, "Proposed Agreement between Robert Beda and Friends of the Land," box 11.2, Ollie Fink to Robert Beda, Sept. 8, 1958, box 71, and Louis B. Seltzer to James Cagney, box 13, all in FOTL.

5. "The Louis Bromfield Institute at Malabar Farm," box 139, folder 2263, and Jonathan Forman to Robert Beda, Dec. 16, 1957, box 144, folder 2309, both in LBC; "Malabar to be First Ecological Center," *MNJ,* Dec. 12, 1957.

6. Floyd B. Chapman to Paul B. Sears, Jan. 7, 1958, and Ollie Fink to Paul Sears, Feb. 14, 1958, both in box 40, folder 606, SP; "Executive Hired for Malabar," *MNJ,* Feb. 11, 1958; "Chapman," *CD,* May 31, 1984.

7. Jonathan Forman, "Epilogue," *LW* 4, no. 3 (Autumn 1958): 23–24; "Trail Site Ideal at Malabar," *MNJ,* June 15, 1958; Floyd B. Chapman, "The Green Gate," *LW* 4, no. 4 (Winter 1958): 8–9. *Land and Water* 4, nos. 2 and 3 (Summer and Autumn 1958) were entirely devoted to these colloquium proceedings.

8. "The Friends of the Land Announce the Opening of Malabar Vegetable Market . . . ," *MNJ,* May 29, 1958; Floyd Chapman, Items Discussed on Inspection of Malabar Farm on March 28, 1958, box 144, folder 2306, LBC; Pete

Cooley to Ollie Fink, June 2, 1958, box 17, FOTL; Schultz and Baumer, *Ohio Dairy Industry;* Baumer et al., *Changing Market Conditions,* 22–44.

9. List of livestock at Malabar Farm, Mar. 31, 1957, box 128, folder 2107, and gross and net receipts from milk sales, 1959, box 140, folder 2265, both in LBC; Petersen and Field, *Dairy Farming,* 385–491; Baumer and Carley, *Bulk Milk Tanks;* Ollie Fink to James H. Hoffman, Oct. 31, 1958, box 11.2, and Ralph Cobey and Richard C. Clark to Holstein Breeders, Sept. 15, 1958, box 16, both in FOTL, Floyd Chapman, *MFN,* no. 1, Mar. 1959.

10. Malabar Farm Project Income and Expense Statements, Year Ended Aug. 31, 1958, and Minutes of the 18th Annual Meeting of the Board of Directors of Friends of the Land, Sept. 19, 1958, both in NF; James Hoffman to Ollie Fink, Nov. 11, 1958, box 11.2, and Ollie Fink, undated manuscript beginning "During the past week an important member . . . ," box 109, both in FOTL.

11. "Listed in 'Who's Who,'" *BTF,* Dec. 27, 1956; "Galion Industrialist Wins State Agricultural Honor," *MNJ,* Aug. 15, 1999; Terricha Bradley, "Galion Grad Gave Life to Community," *BTF,* Jan. 23, 2009; List of Equipment on Loan to Malabar Farm (as at Death of Louis Bromfield, Mar. 1956), box 128, folder 2107, LBC.

12. Catherine Rorick, "They've Worked to Keep Malabar Alive," *MNJ,* May 11, 1957; Anne Rimmer to Chauncey Belknap, Feb. 11, 1957, and Chauncey Belknap to W. M. Alexander, Feb. 14, 1957, both in box 128, folder 2107, LBC; "Malabar Farm Trust Agreement," May 18, 1957, and Bill Solomon to Hope and Bob Stevens, Nov. 14, 1958, both in NF; "Friends of Land Elect," *ABJ,* Sept. 21, 1958.

13. Bill Solomon to Ellen Bromfield Geld, Jan. 17, 1959, MF; Ralph Cobey and Jonathan Forman, resolution of transfer of Malabar to the Louis Bromfield Malabar Farm Foundation, Dec. 8, 1958, FOTL Box 109; Ralph Cobey to Sam Noble, Dec. 10, 1958, and Articles of Incorporation for the Louis Bromfield Malabar Farm Foundation, Dec. 17, 1958, both in NF; "New Group to Acquire Malabar," *MNJ,* Dec. 19, 1958.

14. Ollie Fink to Pete Cooley, Dec. 18, 1958, Ollie Fink to Cecil Forbes, Dec. 22, 1958, Ollie Fink, "Special Report to the Samuel Roberts Noble Foundation," Jan. 12, 1959, and Ollie Fink to Cecil Forbes, n.d., all in box 17, FOTL; Bill Solomon to Ellen Bromfield Geld, Jan. 17, 1959, MF.

15. Jonathan Forman to Ollie Fink, Jan. 15, 1959, Box 109, FOTL.

16. Ollie Fink to Jonathan Forman, n.d., box 109, Ollie Fink to Pete E. Cooley, Mar. 18, 1959, box 17, Ollie Fink, Friends of the Land Executive Committee Meeting minutes, Apr. 10, 1959, box 109; Ollie Fink, Special Report to the Board of Directors, box 109; and Bryce Browning to Ollie Fink, Feb. 16, 1959, box 109, all in FOTL.

17. Ollie Fink to Bryce Browning, Feb. 18, 1959, box 109, Ollie Fink to The Washington News Co., Mar. 23, 1961, box 16, Russell Lord, personal memorandum, Mar. 5, 1960, box 109, and Dudley C. Smith to Ollie Fink, Sept. 4, 1959, box 109, all in FOTL.

18. Dawn Merritt, "The Roaring 20s: A Call to Action," *Outdoor America,* Winter 2012, 25–33; Dawn Merritt, "From the Jazz Age to World War II," *Outdoor America,* no. 2, 2012, 12–21; Dawn Merritt, "From the Fast-Moving Fifties to the 'Sensible' Sixties," *Outdoor America,* no. 3, 2012, 12–23; Preston Bradley, "Thirty-Five Years," and William H. Pringle, "A Banner Year," both in *Outdoor America* 22, no. 1 (Jan.-Feb. 1957): 3–5, 10–13; "Meet the Izaak Walton League," promotional pamphlet, box 109, FOTL; J. W. Penfold, "Save Our Shorelines!"

*Outdoor America* 24, no. 9 (Sept. 1959): 5–8; Frank Gregg. "SOS: A Program for 1960," *Outdoor America* 24, no. 10 (Oct. 1959): 5–8.

19. Frank Gregg to J. W. Penfold, Dec. 10, 1959, Charles F. Schnee, "Friends of the Land," undated manuscript, and Alden J. Erskine to Frank Gregg, Feb. 15, 1960, all in box 10, folder 3, IWLA; George F. Jackson to Friends of the Land, Feb. 24, 1960, and Ollie Fink to Friends of the Land, Feb. 29, 1960, both in box 109, FOTL.

20. E. V. Jotter to Ollie Fink, Mar. 2, 1960, Ray W. Jones to Ollie Fink, Mar. 15, 1960, David W. Dresbeck to Ollie Fink, Mar. 18, 1960, Philip J. Stone to Ollie Fink, Mar. 20, 1960, and Frank Gregg to Ollie Fink, Mar. 24, 1960, all in box 109, FOTL.

21. Memorandum of Understanding between the Izaak Walton League and Friends of the Land, Mar. 24, 1960, Certificate of Dissolution of Friends of the Land, June 10, 1960, and United States District Court for the District of Columbia, order for dissolution of Friends of the Land, Civil Action no. 3300–60, Feb. 20, 1961, all in box 109, FOTL; "Friends of the Land Combines with the Izaak Walton League," *Outdoor America* 26, no. 4 (Apr. 1961): 5; William A. Riaski, "'Friends of the Land' Join the Izaak Walton League," *Outdoor America* 26, no. 5 (May 1961): 4–5.

22. William Riaski to Burt Brickner, Mar. 29, 1961, and William Riaski to Executive Board members, Apr. 6, 1961, both in box 10, folder 3, IWLA; "Fink Papers Acquired by Historians," *ZTR,* July 24, 1971; Ollie Fink to Pete Cooley, Aug. 3, 1959, box 17, FOTL; Burns Harlan, "Of People and the Times," *ZTR,* Feb. 14, 1965; "Ollie Fink Dies at 71," *ZTR,* Mar. 1, 1970; "Man with a Mission," *ZTR,* Mar. 4, 1970.

23. Ollie Fink to Pete Cooley, Apr. 16, 1962, box 17, Russell Lord, "A Personal Memorandum," Mar. 5, 1960, box 109, and Philip J. Stone to Ollie Fink, Mar. 20, 1960, box 109, all in FOTL.

24. Louis Bromfield to Jonathan Forman, circa 1952, box 13, FOTL.

25. "Conservation in the 'Space Age,'" *Outdoor America* 23, no. 4 (Apr. 1958): 15; Paul B. Sears, "The Inexorable Problem of Space," in Shepard and McKinley, *Subversive Science,* 77–93; Hugh Bennett, "Up Stream Flood Control" and "Principles of Permanent Soil Conservation," *Hugh Bennett Lectures,* 35, 52; Udall, *Quiet Crisis,* 245–47; Frank Gregg, "A Long Way from Home," *Outdoor America* 23, no. 2 (1958): 2; Frank Graham Jr., *Since Silent Spring,* 26–29, 71, 167–68; Hazlett, "Story of *Silent Spring,*" 35–36, 179, 384; Wickenden, *Our Daily Poison,* 89; Carson, *Lost Woods,* 99–100; Case, *Organic Profit,* 130–32.

26. Reilly, "Poisoning of Our Drinking Water," 10–23; Carstairs and Elder, "Expertise, Health, and Public Opinion"; Exner and Waldbott, *American Fluoridation Experiment;* Drake, *Loving Nature,* 54–77; "'Fluoridation? Doctors Present Opposing View on Treatment of Columbus Water," *CD,* Feb. 23, 1956; "Allergist, Dentist Argue on Fluoride," *CD,* Mar. 19, 1956; "Election at a Glance," *CD,* May 7, 1958; "Heavy Primary Voting Sets Up Party Tickets," *CD,* May 7, 1958; Charles Nelson, "Proceedings of the Council"; "Dr. Forman Quits OSMJ Editorship," *CD,* Nov. 13, 1958; "Medical Group Fires Dr. Forman as Editor," undated newspaper clipping, box 84, FOTL; "Fluoride Flowing to Water Taps," *CD,* Jan. 17, 1973.

27. McIntosh, *Background of Ecology,* 193–204; Michael G. Barbour, "Ecological Fragmentation in the Fifties," in Cronon, *Uncommon Ground,* 233–55; Worster, *Nature's Economy,* 234–38; Gleason, "Individualistic Concept"; McIntosh, "H. A. Gleason"; Tansley, "Use and Abuse"; Odum, *Fundamentals of Ecology;*

Golley, *Ecosystem Concept;* Odum, "New Ecology"; Susanna B. Hecht, "The Evolution of Agroecological Thought," in Altieri, *Agroecology,* 8–10; Francis et al., "Agroecology," 104–9; Gliessman, *Agroecology,* 18–19.

## 9. Surviving the Sixties

1. Undated transcript of "Dr. Jonathan Forman's Tour of the Big House, Malabar," box 3, folder 12, Floyd Chapman Papers.

2. Floyd Chapman, *MFN* nos. 4–6, 19, July–Sept. 1959, Oct. 1960; Velita Kinney, "Louis Bromfield Would Approve Changes Effected at Malabar," *MNJ,* Aug. 2, 1959; Jonathan Forman to Ralph Cobey, Oct. 15, 1958, box 143, folder 2304, LBC; Fox, Baumer, and Kepner, *Sale of Milk.*

3. Floyd Chapman, *MFN,* nos. 3–6, 19, and 30, June–Sept. 1959, Oct. 1960, and Sept. 1961; "Teacher's Pre-Field Trip Discussion Guide," box 131, folder 2129, and Floyd Chapman, Louis Bromfield Malabar Farm Foundation Report of the Director, Calendar Year 1960, box 143, folder 2304, both in LBC.

4. Floyd Chapman, *MFN,* nos. 2 and 6, May and Sept. 1959; Kinney, "Louis Bromfield"; Harold Friar, *MFN,* no. 15, June 1970; Miraglia, "Seeds of Knowledge," 22–27.

5. Floyd Chapman, *MFN,* nos. 1, 12, and 40, Mar. 1959, Mar. 1960, and July 1962; Dick Collier, "Scientists Call for 'Cold War,'" *MNJ,* Feb. 16, 1959; "Institute to be Held at Malabar, Kingwood," *MNJ,* June 14, 1959; "Cancel Soils Conference," *MNJ,* June 25, 1959; Dick Collier, "Link Good Soil to Health," *MNJ,* June 20, 1960; "Praises Conservation Work of Bromfield," *MNJ,* July 16, 1962.

6. "No Money Troubles, Malabar Board Says," *MNJ,* Sept. 28, 1962; "Professor to Head Malabar Farm Board," *MNJ,* Nov. 20, 1959; Malabar Farm Agricultural Advisory Board meeting minutes, Jan. 9, 1960, May 21, 1960, and Oct. 29, 1960, box 142, folder 2296, LBC; Dick Collier, "Plans Call for Theater in Malabar's 'Big House' Barn," *MNJ,* July 17, 1960; Chapman, Report of the Director, Calendar Year 1960.

7. Floyd Chapman, *MFN,* nos. 23, 26, 28, and 31, Feb., May, July, and Oct. 1961; "A Friendly Link with Africa," *MNJ,* July 7, 1961; "Pupils to Study at Farm," *MNJ,* Apr. 4, 1961; "Malabar 'Junior Explorers' Form," *MNJ,* Apr. 6, 1961; "Camp Fire Girls to Visit Bromfield Home, Hike over Malabar's Trails," *MNJ,* May 24, 1962. The mural that schoolchildren gave the foundation is stored in Bromfield's office at Malabar.

8. Floyd Chapman, *MFN,* nos. 28–30, July-Sept. 1961; Floyd Chapman, "The Louis Bromfield Malabar Farm Foundation Report of the Director, Calendar Year 1961," Jan. 16, 1962, box 139, folder 2262, LBC; "The Louis Bromfield International School of Practical Agriculture," box 2, MF; Virginia Lee, "Fair Impresses Iranian Ambassador," *MNJ,* Aug. 9, 1961; Virginia Lee, "Iranian Envoy Asks Better Understanding," *MNJ,* Aug. 10, 1961.

9. Dick Collier, "Bromfield's Views on Role of Farm Come Alive Again in Malabar School," *MNJ,* Sept. 7, 1961; Floyd Chapman, *MFN,* no. 31, Oct. 1961; Chapman, Report of the Director, Calendar Year 1961; Robert S. Ricksecker to Cecil Forbes, May 24, 1961, Robert S. Ricksecker to Guy H. Woodward, Sept. 7, 1961, and Guy H. Woodward to Robert S. Ricksecker, Sept. 12, 1961, all in NF.

10. Floyd Chapman, *MFN,* no. 8, 11, Nov. 1959, Feb. 1960; Chapman, Report of the Director, Calendar Year 1960; Floyd Chapman to Friend of Malabar Farm,

Apr. 1, 1961, box 142, folder 2294, and John E. Stark, Report of Activities since October 1961, Dec. 18, 1961, box 144, folder 2308, both in LBC; "The Louis Bromfield Farm Foundation," *MNJ,* Dec. 15, 1961; "'Malabar Talks' Slated at Library," *MNJ,* Mar. 17, 1962; "Stark Will Leave Malabar," *MNJ,* June 5, 1962.

11. "'KaRoy' Milk Herd Takes Top Honors," *MNJ,* Sept. 29, 1959; DHIA Report for April 1961, box 133, folder 2154, and Malabar Farm Advisory Board meeting minutes, Apr. 8, 1961 and Mar. 31, 1962, box 142, folder 2296, both in LBC; Chapman, Report of the Director, Calendar Year 1961.

12. Baumer et al., *Changing Market Conditions,* 2–3; Gooding, Baumer, and Eickhoff, *Sale of Dairy Products,* 3; Baumer et al., *Dimensions of Consumer Attitude;* Mitchell and Baumer, *Interim Report,* 6; Jacobson and Hoddick, *Dairy Marketing Cooperatives,* 2, 15; Roy Benson to Member, Northwestern Cooperative Sales Association, Feb. 6, 1962, box 133, folder 2155, LBC.

13. Petersen and Field, *Dairy Farming,* 263–67, 248, 282; Smith-Howard, *Pure and Modern Milk,* 123–46; Bookchin, *Our Synthetic Environment,* 182–87; Jundt, *Greening the Red,* 90–100, 152.

14. "Malabar's Chief Quits," *MNJ,* Sept. 14, 1962; "Not a Happy Tale," *MNJ,* Sept. 17, 1962; "No Money Troubles, Malabar Board Says," *MNJ,* Sept. 28, 1962; "Chapman Takes Post," *MNJ,* Sept. 21, 1962; "Over 3,000 Visit New Flower Store," *MNJ,* Dec. 2, 1962; "Idyllic Stream Saved," *CD,* Nov. 2, 1969; "Gardeners," *CD,* June 9, 1974; Stephen Berry, "Ecologist Trades Trails for Inniswood Gardens," *CD,* June 27, 1978.

15. "Not a Happy Tale"; Jonathan Forman to Ken Woodman, Sept. 19, 1962, MF; "No Money Troubles."

16. "Short Named Director of Foundation," *Hillsboro (OH) Press-Gazette,* May 21, 1963; George N. Constable, "This Week's Personality," *MNJ,* July 19, 1970; "Ex-Game Official's Rites Set," *CD,* Feb. 1, 1975.

17. "Agnes M. Schwartz," *MNJ,* Apr. 5, 2008; "E. B. Howard Named as Malabar Farm Trustee," *MNJ,* Nov. 25, 1961; Regine Kimberly, "Little Known Facts about Well Known People," *ZTR,* July 23, 1961; Bud Harsh, "Success Formula," *ZTR,* Feb. 10, 1963; "Inn Owner, Friend of Fair Dies," *CD,* July 21, 1964; Fern Sharp, "New Inn Is Open for Institute at Malabar," *CD,* July 10, 1963.

18. Sharp, "New Inn"; "Top Scientists Due at Malabar," *MNJ,* July 12, 1963; "Winner of Bromfield Medal Fears Fight to Conserve Soil May Fail," *MNJ,* July 14, 1963; Carson, *Silent Spring,* 12. For more on Rachel Carson and *Silent Spring,* see Lear, *Rachel Carson;* Souder, *Farther Shore;* Brooks, *House of Life,* Frank Graham Jr., *Since Silent Spring,* Hazlett, "Story of *Silent Spring,*" and Dunlap, *DDT.*

19. "Help Offered by Malabar Farm Cited by Director," *MNJ,* July 15, 1963; A. W. Short, *MFN,* nos. 54 and 57, Jan. and Apr. 1964.

20. Floyd Chapman, *MFN,* no. 18, Sept. 1960; Bill Zipf, "Louis Bromfield's Malabar Has Continuing 'Explays,'" *CD,* Sept. 29, 1963; A. W. Short, *MFN,* no. 56, Mar. 1964; "Newsmen Invited to Malabar," *MNJ,* Sept. 13, 1964; "Oxroast Set for Press," *MNJ,* Oct. 12, 1963; A. W. Short to Jonathan Forman, July 21, 1964, filing cabinet, MF; "Doctor Cites Threat to Health in Chemicals," *MNJ,* Aug. 30, 1965; "2-Day Institute Set at Malabar," *MNJ,* June 13, 1968.

21. "E. B. Howard Dies," *ZTR,* July 22, 1964; "Taken to Hospital," *MNJ,* Oct. 14, 1964; "Malabar's Busy Place," *MNJ,* May 29, 1966; A. W. Short to Ernest E. Wooden, Dec. 9, 1966, and A. W. Short to Ralph Cobey, June 10, 1965, both in MF; Harold Friar, *MFN* no. 1, Apr. 1969.

22. A. W. Short, *MFN* no. 67, June 1965; Sandra Fraley, "'Home Folk' Keep Alive Tradition of Malabar Inn," *MNJ,* Apr. 2, 1972; "Bromfield Agitated U.S. Farm Policy," *Newark (OH) Advocate,* July 5, 1975; "Grasshopper Pie Is Inn's Secret," *CPD,* July 16, 1976; "Pauline M. 'Polly' Kunkle Wurtz," *MNJ,* Apr. 17, 2004.

23. "Malabar: Love for Land," *Wonderful World of Ohio,* May 1966, 15–17, box 131, folder 2133, and Tom Thompson, "Malabar Revisited," *Minutes: Magazine of Nationwide Insurance* (Oct. 1968): 17–21, box 121, folder 2025, both in LBC; A. W. Short to Ralph Cobey, Sept. 1, 1967, filing cabinet, MF; Virgil A. Stanfield, "Bromfield's Books in Demand," *MNJ,* Sept. 3, 1967; Harold Friar, *MFN* nos. 3–4, 8, June–July, Nov. 1969; "They Still Like Malabar," *MNJ,* Sept. 4, 1969.

24. Robert Smith, *Ecology of Man;* Worster, *Nature's Economy,* 359–60; Shabecoff, *Fierce Green Fire,* 103–11; Udall, *Quiet Crisis,* 243–44; Dunlap and Mertig, "Evolution of the U.S. Environmental Movement," 210; Kline, *First Along the River,* 79–81; Rothman, *Greening of a Nation?* 115–21; Jundt, *Greening the Red,* 181–216; David Stradling, "Earth Year," and Gaylord Nelson, "Earth Day Speech," in Stradling, *Environmental Moment,* 59, 85–86; Sampson, *Love of the Land,* 200–232.

25. Ellen McClarran, "Bromfield Blames Man for Desolation," *MNJ,* Apr. 13, 1970; Ellen McClarran, "Bromfield Warned of DDT," *MNJ,* Apr. 14, 1970; Robert Batz, "He Preached Ecology When the Word Was Hardly Used," *Dayton (OH) Daily News,* Aug. 3, 1970; George DeVault, "What Became of Walnut Acres?" *Natural Farmer,* Spring 2006, 29–34; personal communications from Rich Collins (Nov. 30, 2018), John Smith (Feb. 3, 2019), Joan Donaldson (July 28, 2019), and Franklin Espriella (Jan. 3, 2020).

26. Harold Friar, *MFN* no. 15, June 1970; "Cobey on Environment Task Force," *BTF,* Mar. 4, 1971; Linda Decker, "Malabar Farm Widens Scope, Gets New Name," *MNJ,* Mar. 10, 1971; "Louis Bromfield Malabar Farm Foundation Profit and Loss Statement," 1969, NF.

27. Faulk, Faulk, and Gray, *Imagination and Ability,* 276; Noble Foundation, *Tracing Our Steps,* 21; Glen W. Allen to James E. Thompson, Mar. 15, 1971, Ralph Cobey to Technical Institute District, May 7, 1971, Ralph Cobey to James Thompson, May 14, 1971, John March to Ralph Cobey, May 28, 1971, James E. Thompson to Ralph Cobey, Sept. 9, 1971, James E. Thompson to Ralph Cobey, Sept. 17, 1971, and Carlton E. Spitzer to John March, Nov. 7, 1971, all in NF.

28. Marguerite Miller, "Foreclosure Threatens Malabar Farm," *MNJ,* Sept. 22, 1971; Ralph Cobey to Sam Noble, Sept. 27, 1971, and Sam Noble to Ralph Cobey, Sept. 29, 1971, both in NF.

### 10. For the People of Ohio

1. William B. Nye to John March, Sept. 27, 1971, and D. K. Woodman to John March, Oct. 1, 1971, both in NF; Bob Hanusz, "Malabar Farm Fighting Off 'Vultures,'" *MNJ,* May 21, 1972; Tom Walton, "Beautiful Malabar Farm," *Toledo (OH) Blade,* May 14, 1972.

2. "The Committee to Save Malabar," box 141, folder 2281, and Senate Bill No. S-72–006, Kent State University, Apr. 19, 1972, box 139, folder 2263, both in LBC.

3. Lila Wagner, "We Want to Help Malabar Farm," *ABJ,* May 11, 1972; Charles Lally, "Children Are Confident Malabar Can Be Saved," *ABJ,* May 28, 1972; Lila Wagner and Ruth Colton, "Louis Bromfield's Malabar Farm from Mecca to State

Park (1956–1972): The Story of the Children's Crusade," brochure, May 3, 1981, 1990–1999 Magazines/Booklets, MF.

4. Walton, "Beautiful Malabar Farm"; Charles Lally, "Foundation-State Deal May Save Malabar," *ABJ,* June 14, 1972; Charles Lally, "State Agrees to Preserve Bromfield's Malabar Farm," *ABJ,* June 15, 1972; Virgil A. Stanfield, "Foundations' Gift to People of Ohio," *MNJ,* June 15, 1972; Warranty Deed, Aug. 15, 1972, Statement of Satisfaction, and Ohio Auditor of State, audit of ODNR's operation of Malabar Farm from July 1, 1972 to Nov. 2, 1975, approved June 23, 1977, box 13, all in MF.

5. "Interim Operating Agreement," Aug. 3, 1972, MF; Ed Kenyon, "State Takes over Malabar," *MNJ,* Aug. 4, 1972; Anne Tifenbach, "Bromfield Would Have Loved It," *MNJ,* Aug. 4, 1972; "Malabar Inn Lease Extended," *MNJ,* July 13, 1972; "Bromfield Agitated U.S. Farm Policy," *Newark (OH) Advocate,* July 5, 1975.

6. Bruce Estes, "Malabar's Farm's Future Plotted during Visit by State Officials," *MNJ,* June 20, 1972; "Soil Laboratory, Children's Zoo Eyed at Malabar," *MNJ,* Aug. 5, 1972; "26 People Asked to Serve as Malabar Farm Advisers," *MNJ,* Aug. 22, 1972; Virgil A. Stanfield, "Malabar Farm Operation to Take $220,000," *MNJ,* Sept. 15, 1972; Virgil Stanfield, "Malabar Farm Advisors Consider Future Uses for Bromfield's Barn," undated newspaper clipping, box 132, folder 2137, LBC.

7. Pat Heydinger, "Malabar Farm 'Curator' Selected," *MNJ,* Nov. 22, 1972; "Bill's Job," *MNJ,* Nov. 25, 1972; "Malabar Farm Can Keep Bromfield Ideas Alive," *MNJ,* Jan. 20, 1973.

8. Dan Kopp, "Malabar Farm to Reflect Life-Style of Bromfield," *MNJ,* Jan. 18, 1973; Jason Thomas, "Ohio Seeks Renewal of Malabar," *CPD,* Sept. 30, 1972; Rodger White, "New 'Top Farmer' Enjoys Malabar," *CD,* Oct. 22, 1972; Tom Brennan, "Restoring Malabar Farm," *MNJ,* Mar. 18, 1973.

9. Wilbur Dunbar, "State Reviving Malabar as Heritage and Showplace," *ABJ,* Jan. 21, 1973; Tom Brennan, "Malabar Park Freed from Wild Bushes," *MNJ,* Jan. 31, 1973; Ed Kenyon, "Bromfield's Trench Silo Covered Up," *MNJ,* July 10, 1973.

10. "Louis Bromfield's Malabar Farm," state brochure, box 2, MF; Lance Wynn, "Architect Submits List of 'Missing' Malabar Items," *MNJ,* July 23, 1974; "State Closes on Alleged Missing Items from Malabar Farm," *MNJ,* Sept. 24, 1974; "Lawmaker Calls Malabar Land Deal 'Shady,' State Official Defends It," *MNJ,* July 19, 1974; "Public Letters," *MNJ,* Sept. 21, 1974; "Malabar Employee Is Facing Deadline," *CD,* May 17, 1977; "Court Will Decide if Sale Legal," *CD,* May 26, 1977.

11. Frank Benson, "Jury Probe of Malabar Asked," *MNJ,* Feb. 21, 1975; "Ex-Malabar Manager Indicted," *MNJ,* Mar. 4, 1975; "Ex-Malabar Aide Pleads Innocent," *MNJ,* Mar. 7, 1975; "Malabar Farm Ex-Chief Is Told to Repay State," *CPD,* Aug. 13, 1975; "Former Malabar Manager Admits Guilt," *MNJ,* Aug. 23, 1975.

12. "Malabar Operation 'Mess,'" *MNJ,* Apr. 24, 1975; "Management of Malabar Is Changed," *CD,* Feb. 5, 1976; Ron Rutti, "Neighbors Reluctant to Give up Farms," *MNJ,* June 22, 1976; "Land Bought for Malabar," *MNJ,* Aug. 3, 1976; Tom Brennan, "Board Turns Down Funds for Malabar Farm," *MNJ,* Sept. 28, 1976; Tom Brennan, "State Admits Malabar Probe," *MNJ,* Oct. 28, 1976; "Board Oks Buying 198 Acres to Extend Malabar Farm Park," *CD,* Nov. 23, 1976; Ohio Auditor of State, 1977 audit of Malabar Farm, 13.

13. "Berry Is Manager at Malabar Farm," *CD,* Sept. 25, 1976; Tom Brennan, "Malabar Chief Looks Ahead," *MNJ,* Oct. 1, 1976; Jim Berry, telephone interview

by Anneliese Abbott, May 7, 2020. For more on living history in the 1970s, see Jay Anderson, *Time Machines;* Kelsey, "Outdoor Museums"; Rymsza-Pawlowska, *History Comes Alive;* and Woods, "Living Historical Farming."

14. Berry interview; Brennan, "Malabar Chief Looks Ahead"; Tom Brennan, "Malabar Aims Programs to Public," *MNJ,* June 11, 1978; Karen Palmer, "Malabar Farm Plans Expansion," *MNJ,* May 20, 1979.

15. Berry interview; Tom Brennan, "'Ohio Heritage Days' to Offer Preview of Malabar Farm Plans," *MNJ,* Sept. 23, 1976; "ODNR Schedules 'Ohio Heritage Days,'" *Logan (OH) Daily News,* Oct. 22, 1976; Tom Brennan, "Trip Back in Time Offered," *MNJ,* Nov. 4, 1976; George N. Constable, "Heritage Days Show Way It Was," *MNJ,* Nov. 7, 1976.

16. "Heritage Days Set at Malabar," *BTF,* Sept. 19, 1977; "Farm a Fun Place," *MNJ,* Sept. 27, 1977; Karen Palmer, "Even Smells Were Old-Fashioned," *MNJ,* Oct. 2, 1977; "Old Times Return to Malabar Farm," *MNJ,* Sept. 20, 1981; Norma Steele, "Heritage Days Draws Crowds," *MNJ,* Sept. 26, 1982.

17. "Malabar Farm Day Enjoyed by Hundreds," *MNJ,* July 30, 1978; "'Farm Day' Coming Up at Malabar," *MNJ,* July 16, 1980; Berry interview; "Plow Show Planned at State Park," *MS,* Apr. 13, 1979; Lee Stratton, "Horsepower Beats Tractor Power in Times of Rising Gasoline Prices," *CD,* Apr. 20, 1979; "Horses Will Plow at Malabar," *CPD,* May 2, 1980.

18. Bundy, *World Economic Crisis;* Karen Merrill, *Oil Crisis;* Wellum, "Energizing the Right"; Jacobs, *Panic at the Pump;* Meadows et al., *Limits to Growth,* 123–26; Agnew, *Back from the Land,* 7–29; Dana Brown, *Back to the Land,* 202–20; Nearing and Nearing, *Living the Good Life.*

19. Fred J. Cook, "Gasohol—A 100-Proof Solution," *Nation,* Apr. 21, 1979, front cover, 432–36; Eliot Janeway, "Gasohol: Solution to the Gas Shortage," *Atlantic,* Nov. 1979, 62–66; "Malabar to Show Off Gasohol Still," *MNJ,* Nov. 7, 1979; Richard G. Ellers, "Fuel for Thought," *CPD,* Jan. 13, 1980; Berry interview.

20. "Wood Heat Day Planned," *MNJ,* Oct. 27, 1978; "State Parks," *CD,* Oct. 26, 1980; "Last Word on Wood," *MNJ,* Oct. 29, 1981; Berry interview.

21. Joan Brown, "Bromfield, Friends Enjoyed Cabin Visits," *MNJ,* Apr. 21, 1974; Bachelder, *These Thousand Acres,* 143–46; Brennan, "Board Turns Down Funds"; "Ohio Division of Parks to Offer Wide Variety of Workshops in 1981," *CD,* Dec. 14, 1980; "Malabar Farm Hosts Workshop and Clinic," *BTF,* Mar. 27, 1984.

22. Berry interview; Tom Brennan, "Sugar Festival Goes without Sap," *MNJ,* Mar. 5, 1978; Ron Simon, "Sweetness," *MNJ,* Mar. 15, 1981; Jim Berry, personal communication, June 26, 2020; Brent Charette, telephone interview by Anneliese Abbott, July 15, 2020.

23. John Veverka et al., "Malabar Farm: An Interpretive Planning Process," report by School of Natural Resources, OSU, 1977, box 10, MF; Ohio Auditor of State, 1977 audit of Malabar Farm; "'Number 51' Very Content Being a Top Milk Producer," *MS,* July 12, 1977; Berry interview.

24. Tom Brennan, "Emphasis Changes to Farm," *MNJ,* May 4, 1977; Brennan, "Malabar Aims Programs to Public"; Fran Murphey, "Dog's Just One of the New Things at Malabar Farm," *ABJ,* Sept. 28, 1980.

25. USDA, *Save Fuel;* Batie, *Soil Erosion,* 5–7; Sampson, *Farmland or Wasteland,* 57–60; William Hayes, *Minimum Tillage Farming,* 24–26; American Farmland Trust, *Soil Conservation in America,* 21–23, 32–38, 50–52; USDA, *2012 National Resources Inventory,* 2–6; Hugh Bennett, *Soil Conservation,* 7–9.

26. American Farmland Trust, *Soil Conservation in America,* xxii; USDA, *Tillage Options;* Little, *Green Fields Forever,* 66–72; Nancy Baker, *Tillage Practices,* 3; Batie, *Soil Erosion,* 44–51; USDA, *2012 National Resources Inventory,* 2–6; Conacher and Conacher, *Herbicides in Agriculture.*

27. "Malabar Will Open Trail," *MNJ,* May 22, 1978; "Trail Dedicated," *MNJ,* Aug. 3, 1978; Jim Underwood, "Rhodes to Open 'New' Malabar," *MNJ,* Sept. 20, 1980; "New Facilities Dedicated at Malabar Farm State Park," *MS,* Sept. 28, 1980; Stuart M. Abbey, "Hostel Opens at Malabar," *CPD,* Jan. 13, 1978.

28. "Malabar Schedules Christmas Tours," *MNJ,* Dec. 11, 1976; Karen Palmer, "Christmastide Still Joyful at Malabar's Big House," *MNJ,* Dec. 14, 1978; Chriss Harris, "Club Decorates Big House at Malabar Farm," *MNJ,* Dec. 20, 1981; "Malabar Farm Offers Lots of Winter Fun," *MNJ,* Jan. 8, 1979; "Malabar Offers Skiing," *Fremont (OH) News-Messenger,* Dec. 11, 1980.

29. "Malabar Inn Equipment to Be Sold," *MNJ,* Oct. 10, 1976; Karen Palmer, "Refurbished Malabar Inn Now Open," *MNJ,* Oct. 22, 1978; "Spring Volksmarch," *MNJ,* Apr. 1, 1981; "'Marsching' at Malabar Farm," *MNJ,* Apr. 12, 1981; "Malabar Farm to Pay Tribute to Bluebirds," *MNJ,* June 10, 1982; "Olde-Tyme Music Fest," *MNJ,* June 21, 1983; "Malabar Film Ready," *ABJ,* May 3, 1981; Wagner and Colton, "Louis Bromfield's Malabar Farm."

30. "Tug of War," *MNJ,* May 12, 1980; "Farmers Have a Field Day," *MNJ,* May 11, 1987; Jacobs, *Panic at the Pump,* 271–93; Wellum, "Energizing the Right," 259–311; Karen Merrill, *Oil Crisis,* 134; Agnew, *Back from the Land,* 89–90, 198–210; Dana Brown, *Back to the Land,* 221–26; "Barnyard Basics on Display," *MNJ,* May 2, 1985.

31. "Malabar Farm Manager Leaving for Cincinnati Post," *MNJ,* May 9, 1986; Berry, personal communication, July 16, 2020; "Lexington Grad Named Park Manager," *MNJ,* Aug. 19, 1986; Charette interview.

### 11. Sustainable Agriculture

1. "Malabar Demo Garden Reaps Impressive Harvest," *Ohio Ecological Food and Farm Association News* 7, no. 4 (1987): 1, and "Malabar Truck Patch Project Revived," *Ohio Ecological Food and Farm Association News* 8, no. 4 (1988): 1, 6, both in OEFFA archives.

2. "The Ohio Ecological Food and Farm Association," brochure, circa 1979, OEFFA archives; Barton, *Global History,* 185; USDA, *Report and Recommendations,* xiv; Youngberg, "Alternative Agricultural Movement," 524; Youngberg and DeMuth, "Organic Agriculture," 299, 305; Francis, "Sustainable Agriculture," 97; Hildebrand, "Agronomy's Role," 286; Keeney, "Sustainable Agriculture."

3. MacRae et al., "Agricultural Science," 184–88; Beus and Dunlap, "Paradigmatic Roots"; Bird, "Sustainable Agriculture Research," 142; Scott Williams, "International Conference Identifies Concerns," *Ohio Ecological Food and Farm Association News* 8, no. 4 (1988): 1, 6, and Food for Peace Movement, "O.S.U. Professors Conspire to Promote Genocide Against Third World, Destruction of U.S. Agriculture," anti-organic flyer, circa 1988, both in OEFFA archives; Beeman and Pritchard, *Green and Permanent Land,* 131, 146; Youngberg and DeMuth, "Organic Agriculture," 316.

4. Scott Williams, personal communication, Apr. 29, 2020; Bromfield, *Louis Bromfield at Malabar;* DeVault, *Return to Pleasant Valley;* Terry Mapes, "Bromfield's Love Letter to Farms in New Edition," *MNJ,* Nov. 2, 1997; Louis

Andres, telephone interview by Anneliese Abbott, Apr. 30, 2020; "Farm Workshop Planned," *MNJ,* July 16, 1988.

5. Brent Charette, telephone interview by Anneliese Abbott, July 15, 2020; "Malabar Farm Hires Rangers," *MNJ,* July 28, 1989.

6. "Malabar Farm to Shut Down Agricultural Activities," *MNJ,* Nov. 8, 1991; Linda Martz, "Malabar Herd Not 'Hamburger,'" *MNJ,* Nov. 20, 1991; Linda Martz, "Malabar Cows Adjust to Transfer to Wooster," *MNJ,* Mar. 11, 1992.

7. Linda Martz, "Bromfield's Granddaughter Fears Malabar Farm Cuts," *MNJ,* Nov. 9, 1991; Linda Mertz, "First Malabar Farm Manager Is Critical of ODNR Park Cuts," *MNJ,* Dec. 22, 1991; Charles H. Frye, "Malabar Farm," *MNJ,* Dec. 27, 1991; Ron Simon, "100-Plus Attend 'Save the Cows' Meeting at Malabar Park," *MNJ,* Nov. 24, 1991; Sheryl Harris and Cheryl Curry, "No Moos, Just Boos at Malabar Farm," *ABJ,* Nov. 24, 1991; Ellen Bromfield Geld, "Cost-Cutters Missed Point of Malabar Farm Park," *MNJ,* Feb. 9, 1992; Annette McCormick, "Save the Cows," *MNJ,* Mar. 21, 1992; Linda Martz, "Group Formed to Return Cows to Malabar," *MNJ,* Feb. 17, 1992; Ron Simon, "What's Malabar without the 'Farm'?" *MNJ,* Mar. 23, 1992; Annette McCormick, telephone interview by Anneliese Abbott, July 13, 2020.

8. Don Baird, "Malabar Fans Protest Moving Cattle." *CD,* Nov. 24, 1991; Glen D. Alexander to Lila S. Wagner, Apr. 28, 1992, NF; Nancy D. Green to George Voinovich, box 13, MF; Charette interview; Sybil Burskey, "Malabar Injustice," *MNJ,* Apr. 19, 1992; Andres interview; "Malabar Gets New Park Manager," *MNJ,* Apr. 11, 1992.

9. Andres interview; Louis Andres, "Malabar Farm Much More than Just a Lot of Cows," *ABJ,* June 5, 1992; Ron Simon, "Ag Director Chats on Mount Jeez," *MNJ,* June 29, 1992; Ron Simon, "Ex-Bromfield Aide Recalls Early Days of Malabar Farm," *MNJ,* Sept. 2, 1992; Linda Martz, "Bromfield Library Found in Basement to Be Rescued," *MNJ,* Mar. 20, 1992; Miraglia, "Seeds of Knowledge"; Karen Palmer, "It's 'Hello Louis' Again," *MNJ,* June 28, 1992; Linda Martz, "Farm Promised 'Special Attention' at Rededication," *MNJ,* Sept. 11, 1992; "Local Business Help with Malabar Farm Projects," *MNJ,* Nov. 1, 1992.

10. Miriam Smith, "Malabar Fund-Raiser for Visitor Center," *MNJ,* Oct. 17, 1990; Miriam Smith, "Governor Pledges Support for Future of Malabar Farm," *MNJ,* Oct. 18, 1990; Linda Martz, "Friends Collect $70,000 for Malabar Farm," *MNJ,* Jan. 16, 1992; "Malabar Farm Board Organized," *MNJ,* Nov. 28, 1992; "Malabar Farm State Park's Malabar 2000: A Plan to Create a Comprehensive Learning Center for the Future," spiral-bound report, May 1994, box 2, MF; Andres interview.

11. Andres interview; Marian Young, "Fire at Malabar," *MNJ,* Apr. 5, 1993; Donna Glenn, "Malabar Barn Lost to Fire," *CD,* Apr. 5, 1993; "Questions Linger after Malabar Fire," *MNJ,* Dec. 18, 1994.

12. Marian Young, "A Time to Rebuild," *MNJ,* Apr. 6, 1993; "Questions Linger"; Dennis Willard, "Malabar Fire Story Changes," *MNJ,* Jan. 23, 1995; Doug Caruso, "Malabar Officials Cleared in Fire," *MNJ,* Mar. 16, 1995; Andres interview.

13. "Getting to Bottom of Malabar Barn Fire," *MNJ,* Dec. 28, 1994; Donna Glenn, "Probe into Malabar Farm Fire in '93 Could Take Several More Weeks," *CD,* Dec. 30, 1994; McCormick interview; Andres interview.

14. Andres interview; Linda Partlow, "State to Expand, Restore Malabar," *MNJ,* July 25, 1993; Donna Glenn, "Barn-Raising to Re-Create Malabar Land-

mark," *CD,* May 1, 1994; "Malabar Farm Set to Raise New Barn," *MNJ,* Sept. 1, 1994; "Malabar Memories," *Timber Framing,* June 1995, 14–17, box 13, MF.

15. Ron Simon, "Malabar Barn Will Rise from Ashes," *MS,* Aug. 11, 1994; "Barn-Raising Set at Malabar This Weekend," undated newspaper clipping, box 13, MF.

16. Rob Pasquinucci, "Barn Raising Today at Malabar," *MNJ,* Sept. 4, 1994; "Malabar Farm Set to Raise New Barn," *MNJ,* Sept. 1, 1994; Ron Simon, "Guild Raises Ruckus and Barn," *MNJ,* Sept. 5, 1994; "Finishing Touches," *MNJ,* Sept. 16, 1994; Andres interview.

17. "Malabar 2000: A Vision for the Future," brochure, 1993, 1990–1999 Magazines/Booklets, MF; "Malabar Farm State Park's Malabar 2000"; Andres interview; Wedin and Fales, *Grassland,* 38–42; Nott, *Evolution of Dairy Grazing,* 3–10; Zartman, *Intensive Grazing.*

18. "Intensive Grazing Field Day Planned at Malabar," newspaper clipping, June 30, 1994, box 13, MF; Marcia Schonberg, "There's Cow Pies Aplenty at Malabar Farm," *MNJ,* Oct. 18, 1996; Andres interview; "ODNR Failing in North Central Ohio," *MNJ,* July 4, 1995.

19. "ODNR Failing"; Allitt, *Climate of Crisis,* 184. For more on the biocentric worldview, see Sears, *This Is Our World;* Leopold, *Sand County Almanac;* Devall and Sessions, *Deep Ecology;* Foreman, *Confessions;* Woodhouse, *Ecocentrists;* and Robert Nelson, *New Holy Wars.*

20. Cronon, *Uncommon Ground,* 85. For more on free-market environmentalism, see Huber, *Hard Green,* and Allitt, *Climate of Crisis.*

21. Andres interview. For more recent debates on fracking and logging on public lands near Malabar, see Lou Whitmire, "Opponents of Expanded Park Drilling Try to Get Word Out in Lucas," *MNJ,* Apr. 22, 2011; Jami Kinton, "Protecting the Parks," *MNJ,* May 18, 2012; Monroe Trombly, "'Significant Hazard,'" *MNJ,* Mar. 15, 2020; David A. Lipstreu, "Muskingum Conservancy Should Set a Better Example," *MNJ,* Apr. 19, 2020.

22. Andres interview; Ron Simon, "Remembering Bromfield, His Inspiration Is Our Best Option," *MNJ,* Oct. 1, 1995.

23. Karen Palmer, "Bromfield's Vegetable Garden, Roadside Stand Started Running Again," *MNJ,* May 5, 1999; Andres interview; Mick Luber, telephone interview by Anneliese Abbott, May 1, 2020.

24. Andres interview; "New Name Given to Malabar Restaurant," *MNJ,* Mar. 7, 2006; Lisa Miller, "Charming French Restaurant Comes Off as Pleasant, Not Pretentious," *MNJ,* Sept. 13, 2007.

25. Andres interview; Mark Caudill, "Concern over Felons Working Malabar Won't Change 'Win-Win' Bargain," *MNJ,* May 7, 2006.

26. Ron Simon, "Malabar Farm Getting $3 Million Education Center," *MNJ,* Feb. 24, 2000; "Malabar Funds Growing," *MNJ,* July 11, 2000; Ron Simon, "Industrialist Honored," *MNJ,* June 24, 2001; David Benson, "Malabar Farm Goes Geothermal," *MNJ,* Feb. 8, 2002; Karen Palmer, "Always the Home of Innovation, Malabar Farms Has Gone Geothermal," *MNJ,* Aug. 23, 2002.

27. Andres interview; Ron Simon, "Malabar Growing Plans for the Future," *MNJ,* Aug. 17, 2005; Ron Simon, "New Center in Works at Malabar Farm," *MNJ,* Sept. 25, 2005; Lisa Miller, "Wind Helps Power Malabar Farm Visitor Center," *MNJ,* July 30, 2007.

28. "It's Getting 'Greener' at Malabar Farm State Park," *MNJ,* Apr. 16, 2006; Ron Simon, "Governor Helps Malabar Launch Its Visitor Center," *MNJ,* Sept. 24,

2006; Ron Simon, "Malabar Farm Visitor's Center Enhances Park Experience," *Coshocton (OH) Tribune,* May 24, 2007; Andres interview.

29. "Olympics Down on the Farm," *MNJ,* Aug. 11, 1994; Dan Kopp, "Malabar Gets Theatrical," *Chillicothe (OH) Gazette,* May 15, 2003; Norm Narvaja, "Get Ready for Malabar's Night Haunt," *MNJ,* Aug. 11, 2005; Lou Whitmire, "Nativity Scene Back Up at Malabar Farm," *MNJ,* Dec. 13, 2007; "Nativity Going Up at Malabar," *MNJ,* Nov. 25, 2008.

30. Lou Whitmire, "Andres Retires from State Parks," *MNJ,* July 22, 2012; Andres interview; Korre Boyer, interview by Anneliese Abbott, July 6, 2015, Malabar Farm; Matt Eiselstein, "ODNR Director Announces New Direction for Malabar Farm State Park," *Targeted News Service,* Nov. 30, 2012; Blanco and Lal, *Soil Conservation,* 201–14; Lal, Reicosky, and Hanson, "Evolution of the Plow"; Lal and Bruce, "World Cropland Soils"; Blevins et al., "Conservation Tillage."

31. Lou Whitmire, "Malabar Makeover Set to Start," *MNJ,* May 11, 2013; Lou Whitmire, "Teachers Repainting Malabar Barn Mural," *MNJ,* July 4, 2013; Thomas Bachelder, interview by Anneliese Abbott, July 6, 2015, Malabar Farm; Ginnie Baker, "Winter Wonderland," *MNJ,* Jan. 17, 2013; Ginnie Baker, "Malabark in the Park Event for Canines and Their Humans," *MNJ,* May 7, 2013; Lou Whitmire, "'Sunday Drive Car Show' at Malabar Farm on Sunday," *MNJ,* July 29, 2016; Seth Weibel, "Malabar Farm Restaurant Historic, Tasty," *MNJ,* Nov. 11, 2012; Seth Weibel, "Historic Malabar Inn Has Everything," *MNJ,* Nov. 6, 2015.

32. McCormick interview; Lou Whitmire, "Malabar Farm State Park Manager Resigns," *MNJ,* Jan. 27, 2018; Lou Whitmire, "Ashland County Woman Is New Malabar Manager," *MNJ,* May 20, 2019; Jennifer Roar, telephone interview by Anneliese Abbott, June 30, 2020.

### Conclusion

1. USDA, *Agricultural Statistics 1952,* 635; "Manifesto," *Land* 1, no. 1 (Winter 1941): 12.

2. USDA Economic Research Service, "Selected Charts from Ag and Food Statistics: Charting the Essentials, February 2020," https://www.ers.usda.gov/webdocs/publications/96957/ap-083.pdf?v=8070.9.

3. Beeman and Pritchard, *Green and Permanent Land,* 131, 146.

4. Bromfield, *Malabar Farm,* 230–31.

5. Bromfield, *From My Experience,* 8–9.

# Bibliography

### Archival Sources

Louis Bromfield Collection, SPEC.RARE.CMS.95. Rare Books and Manuscripts Library, Ohio State Univ., Columbus.
Floyd B. Chapman Papers, MSS 1017. Ohio History Connection, Columbus.
Friends of the Land Records Papers, MSS 364. Ohio History Connection, Columbus.
Jonathan Forman Papers, Spec.201121. Forman. Medical Heritage Center Archives, Ohio State Univ., Columbus.
Izaak Walton League of America Records (Ohio), MSS 882. Ohio History Connection, Columbus.
Malabar Farm Archives. Malabar Farm State Park, Lucas, OH.
Samuel Roberts Noble Foundation Archives. Noble Research Institute, Ardmore, OK.
Ohio Ecological Food and Farm Association archives. Wisconsin Historical Society, Madison.
Sears (Paul Bigelow) Papers, MS663. Manuscripts and Archives, Yale Univ., New Haven.

### Published Sources, Dissertations, and Thesis

Adams, Ansel, and Nancy Newhall. *This Is the American Earth.* San Francisco: Sierra Club, 1960.
Agnew, Elenor. *Back from the Land: How Young Americans Went to Nature in the 1970s, and Why They Came Back.* Chicago: Ivan R. Dee, 2004.
Albrecht, William A. *Soil Fertility and Animal Health: The Albrecht Papers,* Vol. 2. Austin, TX: Acres USA, 2005.
Albrecht, William A., and G. E. Smith. "Biological Assays of Soil Fertility." *Soil Science Society Proceedings* 6 (1941): 252–58.
Alderman, Derek H. "Channing Cope and the Making of a Miracle Vine." *Geographical Review* 94, no. 2 (Apr. 2004): 157–77.
Allen, R. R., and C. R. Fenster. "Stubble-Mulch Equipment for Soil and Water Conservation in the Great Plains." *Journal of Soil and Water Conservation* 41, no. 1 (Jan.–Feb. 1986): 11–16.
Allitt, Patrick. *A Climate of Crisis: America in the Age of Environmentalism.* New York: Penguin, 2014.

Allred, Berten W., and Jeff C. Dykes, eds. *Flat Top Ranch: The Story of a Grassland Venture.* Norman: Univ. of Oklahoma Press, 1957.

Altieri, Miguel A. *Agroecology: The Scientific Basis of Alternative Agriculture.* Boulder, CO: Westview Press, 1987.

American Farmland Trust. *Soil Conservation in America: What Do We Have to Lose?* Washington, DC: American Farmland Trust, 1984.

Amundson, Ronald. "Philosophical Developments in Pedology in the United States: Eugene Hilgard and Milton Whitney." In *Footprints in the Soil: People and Ideas in Soil History,* edited by Benno P. Warkentin, 149–65. Amsterdam: Elsevier, 2006, 149–65.

Anderson, David D. *Louis Bromfield.* New York: Twayne, 1964.

Anderson, Chad. "Rediscovering Native North America: Settlements, Maps, and Empires in the Eastern Woodlands." *Early American Studies* 14, no. 3 (Summer 2016): 478–505.

Anderson, Jay. *Time Machines: The World of Living History.* Nashville, TN: American Association for State and Local History, 1984.

Anderson, Joseph L. *Industrializing the Corn Belt: Agriculture, Technology, and Environment, 1945–1972.* DeKalb: Northern Illinois Univ. Press, 2009.

Anderson, Oscar E. *The Health of a Nation: Harvey W. Wiley and the Fight for Pure Food.* Chicago: Univ. of Chicago Press, 1958.

Anderson, Wallace L., and Frank C. Edminster. *The Multiflora Rose for Fences and Wildlife.* USDA Leaflet 374. Washington, DC: GPO, 1954.

Ashby, James. "The Aluminum Legacy: The History of the Metal and Its Role in Architecture." *Construction History* 15 (1999): 80–82.

Bachelder, Thomas. *These Thousand Acres: The Story of the Land That Became Malabar Farm.* Self-published, 2017.

Backer, Kellen. "World War II and the Triumph of Industrialized Food." PhD diss., Univ. of Wisconsin–Madison, 2012.

Bailey, Liberty Hyde. *Liberty Hyde Bailey: Essential Agrarian and Environmental Writings.* Edited by Zachary Michael Jack. Ithaca, NY: Cornell Univ. Press, 2008.

Baker, Nancy T. *Tillage Practices in the Conterminous United States, 1989–2004—Datasets Aggregated by Watershed.* USGS Data Series 573. Reston, VA: US Geological Survey, 2011.

Baker, Richard H., and John I. Falconer. *Costs of Producing Milk in Ohio, 1945–1946.* Research Bulletin 687. Wooster: OAES, 1948.

Banner, Stuart. *How the Indians Lost Their Land: Law and Power on the Frontier.* Cambridge, MA: Harvard Univ. Press, 2005.

Barnes, Robert F., C. Jerry Nelson, Michael Collins, and Kenneth J. Moore. *Forages: An Introduction to Grassland Agriculture,* Vol. 1. 6th ed. Ames: Iowa State Univ. Press, 2003.

Barton, Gregory A. *The Global History of Organic Farming.* Oxford: Oxford Univ. Press, 2018.

Baumer, Elmer F., and Dale H. Carley. *Bulk Milk Tanks on Ohio Farms.* Research Bulletin 776. Wooster: OAES, 1956.

Baumer, Elmer F., Dan I. Padberg, Karl W. Kepner, Thomas A. Klein, and Wesley D. Seitz. *Changing Market Conditions: Implications for Ohio Dairy Marketing Cooperatives.* A. E. 363. Wooster: OAES, 1964.

Baumer, Elmer F., W. K. Brandt, Robert E. Jacobson, and Francis E. Walker. *Dimensions of Consumer Attitude in Fluid Milk Purchases with Special Reference to Doorstep Delivery vs. Store Purchases.* Wooster: OARDC, 1969.

Barton, Gregory A. *The Global History of Organic Farming.* Oxford: Oxford Univ. Press, 2018.

Batie, Sandra S. *Soil Erosion: Crisis in America's Croplands?* Washington, DC: Conservation Foundation, 1983.

Bear, Firman E. "Facts . . . And Fancies about Fertilizer." *Plant Food Journal* 1, no. 2 (Apr.–June 1947): 4–5, 19–23.

Bear, Firman E., Stephen J. Toth, and Arthur L. Prince. "Variation in Mineral Composition of Vegetables." *Soil Science Society Proceedings* (1948): 380–84.

Beatty, Robert O. "The Conservation Movement." *Annals of the American Academy of Political and Social Science* 281, no. 1 (1952): 10–19.

Beeman, Randal S. "The Trash Farmer." *Journal of Sustainable Agriculture* 4, no. 1 (1994): 91–102.

Beeman, Randal S., and James A. Pritchard. *A Green and Permanent Land: Ecology and Agriculture in the Twentieth Century.* Lawrence: Univ. Press of Kansas, 2001.

Beeson, Kenneth C. "The Effect of Mineral Supply on the Mineral Concentration and Nutritional Quality of Plants." *Botanical Review* 12, no. 7 (July 1946): 424–55.

———. *The Mineral Composition of Crops, with Particular Reference to the Soils in Which They Were Grown: A Review and Compilation.* USDA miscellaneous publication 369. Washington, DC: GPO, 1941.

Beeson, Kenneth C., and Gennard Matrone. *The Soil Factor in Nutrition: Animal and Human.* New York: Marcel Dekker, 1976.

Belasco, Warren. "Algae Burgers for a Hungry World? The Rise and Fall of Chlorella Cuisine." *Technology and Culture* 38, no. 3 (July 1997): 608–34.

Bennett, Hugh H. "Adjustment of Agriculture to Its Environment." *Annals of the Association of American Geographers* 33, no. 4 (Dec. 1943): 163–94.

———. *The Hugh Bennett Lectures.* Raleigh: North Carolina State College, Agricultural Foundation, 1959.

———. *Soil Conservation.* New York: McGraw-Hill, 1939.

———. *Soil Erosion a National Menace.* USDA Circular no. 33. Washington, DC: GPO, 1928.

———. *Thomas Jefferson, Soil Conservationist.* USDA Miscellaneous Publication 548. Washington, DC: GPO, 1944.

Bennett, Hugh H., and William Clayton Pryor. *This Land We Defend.* New York: Longmans, Green, 1942.

Bennett, Merrill K. "Population and Food Supply: The Current Scare." *Scientific Monthly* 68, no. 1 (1949): 17–26.

Beus, Curtis E., and Riley E. Dunlap. "Conventional Versus Alternative Agriculture: The Paradigmatic Roots of the Debate." *Rural Sociology* 55, no. 4 (1990): 590–616.

Bird, George W. "Sustainable Agriculture Research and Education Program: With Special Reference to the Science of Ecology." *Journal of Sustainable Agriculture* 2, no. 3 (1992): 141–52.

Biskind, Morton S. "DDT Poisoning and the Elusive 'Virus X': A New Cause for Gastro-Enteritis." *American Journal of Digestive Diseases* 16, no. 3 (Mar. 1949): 79–84.

Biskind, Morton S., and Irving Bieber. "DDT Poisoning: A New Syndrome with Neuropsychiatric Manifestations." *American Journal of Psychotherapy* 3, no. 2 (Apr. 1949): 261–70.

Black, John D. *Food Enough.* Lancaster, PA: Jaques Cattell, 1943.

Blanco, Humberto, and Rattan Lal. *Principles of Soil Conservation and Management.* New York: Springer, 2008.

Blevins, Robert L., Rattan Lal, J. W. Doran, G. W. Langdale, and W. W. Frye. "Conservation Tillage for Erosion Control and Soil Quality." In *Advances in Soil and Water Conservation,* edited by F. J. Pierce and W. W. Frye, 51–68. Chelsea, MI: Sleeping Bear Press, 1998.
Blum, Barton. "Composting and the Roots of Sustainable Agriculture." *Agricultural History* 66, no. 2 (Spring 1992): 171–88.
Bookchin, Murray [Lewis Herber, pseud.]. *Our Synthetic Environment.* New York: Alfred A. Knopf, 1962.
Boyd-Orr, John. *As I Recall: The 1880s to the 1960s.* London: Macgibbon & Kee, 1966.
———. *Food and the People.* London: Pilot Press, 1943.
Brady, Nyle C., and Weil, Ray R. *Elements of the Nature and Properties of Soils,* 3rd ed. Upper Saddle River, NJ: Prentice Hall, 2010.
Brandt, C. Stafford, and Kenneth C. Beeson. "Influence of Organic Fertilization on Certain Nutritive Constituents of Crops." *Soil Science* 7, no. 6 (June 1951): 449–54.
Branyan, Robert L. "From Monopoly to Oligopoly: The Aluminum Industry after World War II." *Southwestern Social Science Quarterly* 43, no. 3 (1962): 242.
Brinkley, Douglas. *Rightful Heritage: Franklin D. Roosevelt and the Land of America.* New York: HarperCollins, 2016.
Brink, Wellington. *Big Hugh: The Father of Soil Conservation.* New York: Macmillan, 1950.
———. "Soil Put to Clinical Tests at Tar Hollow Conference." *Soil Conservation* 8, no. 4 (Oct. 1942): 78–80.
———. "Tar Hollow's Way of Teaching." *Soil Conservation* 9, no. 4 (Oct. 1943): 89–92.
Bristow, William H., and Katherine M. Cook. *Conservation in the Education Program.* Washington: US Department of the Interior, 1937.
Bromfield, Louis. *Early Autumn: The Story of a Lady.* New York: Grosset & Dunlap, 1926.
———. *The Farm.* New York: Harper, 1933.
———. *From My Experience.* 1955. Reprint, Wooster, OH: Wooster Book, 1999.
———. *The Green Bay Tree.* New York: Grosset & Dunlap, 1924.
———. *Louis Bromfield at Malabar: Writings on Farming and Country Life.* Edited by Charles A. Little. Baltimore: Johns Hopkins Univ. Press, 1988.
———. *Malabar Farm.* 1948. Reprint, Wooster, OH: Wooster Book, 1999.
———. *Out of the Earth.* 1950. Reprint, Wooster, OH: Wooster Book, 2003.
———. *Pleasant Valley.* 1945. Reprint, Wooster, OH: Wooster Book, 1999.
———. *The Rains Came: A Novel of Modern India.* New York: Harper, 1937.
———. *The Wealth of the Soil: Soil Is the Fundamental Basis of All Farm Prosperity.* Detroit, MI: Harry Ferguson, 1952.
Brooks, Paul. *The House of Life: Rachel Carson at Work, with Selections from Her Writings Published and Unpublished.* Boston: Houghton Mifflin, 1972.
Brown, Dana. *Back to the Land: The Enduring Dream of Self-Sufficiency in Modern America.* Madison: Univ. of Wisconsin Press, 2011.
Brown, Morrison. *Louis Bromfield and His Books: An Evaluation.* Fair Lawn, NJ: Essential Books, 1957.
Bundy, William P., ed. *The World Economic Crisis.* New York: W. W. Norton, 1975.
Burch, Guy Irving, and Elmer Pendell. *Population Roads to Peace or War.* Washington, DC: Population Reference Bureau, 1945.
Callahan, North. *TVA: Bridge over Troubled Waters.* New York: A. S. Barnes, 1980.
Carpenter, Kenneth J. "A Short History of Nutritional Science: Part 2 (1885–1912)." *Journal of Nutrition* 133, no. 4 (Apr. 2003): 975–84.

———. "A Short History of Nutritional Science: Part 3 (1912–1944)." *Journal of Nutrition* 133, no. 10 (Oct. 2003): 3023–32.
Carson, Rachel. *Lost Woods: The Discovered Writing of Rachel Carson.* Edited by Linda Lear. Boston: Beacon Press, 1998.
———. *Silent Spring.* 1962. Reprint, Boston: Mariner, 2002.
Carstairs, Catherine, and Rachel Elder. "Expertise, Health, and Public Opinion: Debating Water Fluoridation, 1945–80." *Canadian Historical Review* 89, no. 3 (Sept. 2008): 345–71.
Carver, George Washington. *How to Build Up Worn Out Soils.* Bulletin number 6. Tuskegee, AL: Tuskegee Normal and Industrial Institute, 1905.
Case, Andrew N. *The Organic Profit: Rodale and the Making of Marketplace Environmentalism.* Seattle: Univ. of Washington Press, 2018.
Chandler, William U. *The Myth of TVA: Conservation and Development in the Tennessee Valley, 1933–1983.* Cambridge, MA: Ballinger, 1984.
Charles, F. E. "Tar Hollow Conservation-Teaching Laboratory." *Soil Conservation* 6, no. 4 (Oct. 1940): 104, 106.
Chase, Stuart. *Rich Land, Poor Land: A Study of Waste in the Natural Resources of America.* New York: Whittlesey, 1936.
*Chemicals in Food Products: Hearings before the House Select Committee to Investigate the Use of Chemicals in Food Products* (1950). House of Representatives, 81st Congress. Washington, DC: GPO, 1951.
*Chemicals in Food Products: Hearings before the House Select Committee to Investigate the Use of Chemicals in Food Products,* Part 1. House of Representatives, 82nd Congress. Washington, DC: GPO, 1951.
*Chemicals in Food and Cosmetics: Hearings before the House Select Committee to Investigate the Use of Chemicals in Foods and Cosmetics,* Part 2. House of Representatives, 82nd Congress. Washington, DC: GPO, 1952.
Clarke, Nell Ray. "Uncle Sam Raises Bugs." *Scientific American* 138, no. 2 (Feb. 1928): 148–50.
Clements, Frederic E. *Dynamics of Vegetation.* New York: H. W. Wilson, 1949.
———. "Experimental Ecology in the Public Service." *Ecology* 16, no. 3 (July 1935): 342–63.
Clepper, Henry, ed. *Origins of American Conservation.* New York: Ronald, 1966.
Cochrane, Willard W. *The Development of American Agriculture: A Historical Analysis,* 2nd ed. Minneapolis: Univ. of Minnesota Press, 1993.
Commoner, Barry. *The Closing Circle: Nature, Man, and Technology.* New York: Alfred A. Knopf, 1971.
———. *Science and Survival.* New York: Viking Press, 1966.
Conacher, Jeanette, and Arthur Conacher. *Herbicides in Agriculture: Minimum Tillage, Science and Society.* Nedlands, Western Australia: Univ. of Western Australia, Sept. 1986.
Conford, Philip. *The Origins of the Organic Movement.* Edinburgh, UK: Floris Books, 2001.
Conkin, Paul K. *A Revolution Down on the Farm: The Transformation of American Agriculture since 1929.* Lexington: Univ. Press of Kentucky, 2008.
Connelly, Matthew. *Fatal Misconception: The Struggle to Control World Population.* Cambridge, MA: Harvard Univ. Press, 2008.
Conrey, Guy W., J. S. Cutler, and A. H. Paschall. *Soil Erosion in Ohio.* Bulletin 589. Wooster: OAES, 1937.
Cope, Channing. *Front Porch Farmer.* Atlanta, GA: Turner E. Smith, 1949.

Cronon, William. *Changes in the Land: Indians, Colonists, and the Ecology of New England.* New York: Hill & Wang, 1983.

———. "Modes of Prophecy and Production: Placing Nature in History." *Journal of American History* 76, no. 4 (Mar. 1990): 1122–31.

———. "A Place for Stories: Nature, History, and Narrative." *Journal of American History* 78, no. 4 (Mar. 1992): 1347–76.

———, ed. *Uncommon Ground: Toward Reinventing Nature.* New York: W. W. Norton, 1995.

Dale, Tom D., and Grover F. Brown. *Grass Crops in Conservation Farming.* USDA Farmers' Bulletin 2080. Washington, DC: GPO, May 1955.

Davis, Frederick Rowe. *Banned: A History of Pesticides and the Science of Toxicology.* New Haven: Yale Univ. Press, 2014.

Davis, John H., and Ray A. Goldberg. *A Concept of Agribusiness.* Boston: Division of Research, Graduate School of Business Administration, Harvard Univ., 1957.

Davis, John H., and Kenneth Hinshaw. *Farmer in a Business Suit.* New York: Simon & Schuster, 1957.

Deitmeyer, Carl. "Wilfred M. Schutz, Seasoned Conservationist." *Plant Food Journal* 3, no. 4 (Oct.–Dec. 1949): 2–3, 10, 19.

Devall, Bill, and George Sessions. *Deep Ecology: Living as if Nature Mattered.* Salt Lake City: Gibbs Smith, 1985.

DeVault, George, ed. *Return to Pleasant Valley: Louis Bromfield's Best from* Malabar Farm *and His Other Country Classics.* Chillicothe, IL: American Botanist, 1996.

Dillard, Paul. "Chisel Plow Dedicated as ASAE Historic Landmark." *Engineering and Technology for a Sustainable World* 8, no. 1 (Jan. 2001): 21.

Dodd, D. R. *Erosion Control in Ohio Farming.* Agricultural Extension Service Bulletin 186. Columbus: OSU, 1937.

Dodd, Norris E. "The Food and Agriculture Organization of the United Nations: Its History, Organization, and Objectives." *Agricultural History* 23, no. 2 (Apr. 1949): 81–86.

Doordan, Dennis. "Promoting Aluminum: Designers and the American Aluminum Industry." *Design Issues* 9, no. 3 (1993): 45.

Drache, Hiram M. *History of U.S. Agriculture and Its Relevance to Today.* Danville, IL: Interstate Publishers, 1996.

Drake, Brian Allen. *Loving Nature, Fearing the State: Environmentalism and Antigovernment Politics before Reagan.* Seattle: Univ. of Washington Press, 2013.

Duncan, Dayton, Ken Burns, and Julie Dunfey. *The Dust Bowl: An Illustrated History.* San Francisco: Chronicle Books, 2012.

Dunlap, Riley E., and Angela G. Mertig. "The Evolution of the U.S. Environmental Movement from 1970 to 1990: An Overview." *Society and Natural Resources* 4, no. 3 (1991): 209–18.

Dunlap, Thomas R. *DDT: Scientists, Citizens, and Public Policy.* Princeton, NJ: Princeton Univ. Press, 1981.

DuPuis, E. Melanie. *Nature's Perfect Food: How Milk Became America's Drink.* New York: New York Univ. Press, 2002.

Eckelberry, Roscoe H. "A Project in Teacher Education." *Educational Research Bulletin* 23, no. 2 (Feb. 16, 1944): 39–45.

Eckles, Clarence H., and Ernest L. Anthony. *Dairy Cattle and Milk Production,* 4th ed. New York: Macmillan, 1950.

"Educational News and Editorial Comment." *Elementary School Journal* 37, no. 9 (May 1937): 641–52.

Ehrlich, Paul R. *The Population Bomb*. New York: Ballantine, 1968.
Eppig, Margaret L. "Russell Lord and the Permanent Agriculture Movement: An Environmental Biography." PhD diss., Antioch Univ., 2017.
Exner, Frederick B., and George L. Waldbott. *The American Fluoridation Experiment*. New York: Devin-Adair, 1957.
Fanning, Delvins, and Mary Christine Balluff Fanning. "Milton Whitney: Soil Survey Pioneer." *Soil Survey Horizons* 42, no. 3 (Fall 2001): 83–89.
Faulkner, Edward H. *Plowman's Folly*. Norman: Univ. of Oklahoma Press, 1943.
———. *A Second Look*. Norman: Univ. of Oklahoma Press, 1947.
———. *Soil Development*. Norman: Univ. of Oklahoma Press, 1952.
Faulk, Odie B., Laura E. Faulk, and Sally M. Gray. *Imagination and Ability: The Life of Lloyd Noble*. Oklahoma City: Oklahoma Heritage Association, 1995.
Fink, Ollie E. *Conservation for Tomorrow's America*. Columbus: Ohio Division of Conservation and Natural Resources, 1942.
———. "Developing the Program of Conservation Education in Ohio." *Science Education* 25, no. 3 (Mar. 1941): 124–30.
Fite, Gilbert C. *American Agriculture and Farm Policy since 1900*. Washington, DC: American Historical Association, 1964.
Fitzgerald, Deborah. *Every Farm a Factory: The Industrial Ideal in American Agriculture*. New Haven: Yale Univ. Press, 2003.
Flader, Susan L. *Thinking Like a Mountain: Aldo Leopold and the Evolution of an Ecological Attitude toward Deer, Wolves, and Forests*. Columbia: Univ. of Missouri Press, 1974.
Fleming, Donald. "Roots of the New Conservation Movement." *Perspectives in American History* 6 (1972): 7–91.
Foreman, Dave. *Confessions of an Eco-Warrior*. New York: Harmony Books, 1991.
Forman, Jonathan. "How I Came into Medicine." In *Why We Became Doctors*, edited by Noah D. Fabricant, 23–41. New York: Grune & Stratton, 1954.
Forman, Jonathan, and Ollie E. Fink, eds. *Soil, Food and Health: "You Are What You Eat."* Columbus, OH: Friends of the Land, 1948.
———, eds. *Water and Man: A Study in Ecology*. Columbus, OH: Friends of the Land, 1950.
Forseth, Irwin N., and Anne F. Innis. "Kudzu (*Pueraria montana*): History, Physiology, and Ecology Combine to Make a Major Ecosystem Threat." *Critical Reviews in Plant Sciences* 23, no. 5 (2004): 401–13.
Fox, William J., Elmer F. Baumer, and Karl W. Kepner. *The Sale of Milk through Home Dispensers*. Research Circular 100. Wooster: OARDC, Apr. 1961.
Francis, Charles A. "Sustainable Agriculture: Myths and Realities." *Journal of Sustainable Agriculture* 1, no. 1 (1990): 97–106.
Francis, Charles A., et al. "Agroecology: The Ecology of Food Systems." *Journal of Sustainable Agriculture* 22, no. 3 (2003): 99–118.
Fraser, Colin. *Tractor Pioneer: The Life of Harry Ferguson*. Athens: Ohio Univ. Press, 1973.
Funderburk, Robert Steele. *The History of Conservation Education in the United States*. Nashville, TN: George Peabody College for Teachers, 1948.
Gardner, Bruce L. *American Agriculture in the Twentieth Century: How It Flourished and What It Cost*. Cambridge, MA: Harvard Univ. Press, 2002.
Geld, Ellen Bromfield. *The Heritage: A Daughter's Memories of Louis Bromfield*. Athens: Ohio Univ. Press, 1999.
———. *Strangers in the Valley*. New York: Dodd, Mead, 1957.

Gibbard, Stuart. *The Ford Tractor Story Part One: Dearborn to Dagenham 1917–1964.* Ipswich, UK: Old Pond, 1998.

Gleason, Henry A. "The Individualistic Concept of the Plant Association." In *Foundations of Ecology: Classic Papers with Commentaries,* edited by Leslie A. Real and James H. Brown, 98–117. Chicago: University of Chicago Press, 1991.

Gliessman, Stephen R. *Agroecology: The Ecology of Sustainable Food Systems,* 2nd ed. Boca Raton, FL: CRC Press, 2007.

Golley, Frank Benjamin. *A History of the Ecosystem Concept in Ecology: More Than the Sum of the Parts.* New Haven: Yale Univ. Press, 1993.

Gooding, David I., Elmer F. Baumer, and William D. Eickhoff. *Factors Relating to the Sale of Dairy Products in Retail Stores.* Research Bulletin 910. Wooster: OAES, 1962.

Gordon, William E. "Can Traditional Ecology Embrace Human Ecology?" A Letter to the Editors." *Ecology* 31, no. 3 (July 1950): 489–91.

Graham, Edward H. "Ecology and Land Use." *Soil Conservation* 6, no. 5 (Nov. 1940): 123–28.

———. *Natural Principles of Land Use.* New York: Oxford Univ. Press, 1944.

Graham, Frank Jr. *Man's Dominion: The Story of Conservation in America.* New York: M. Evans, 1971.

———. *Since Silent Spring.* Boston: Houghton Mifflin, 1970.

Great Plains Committee. *The Future of the Great Plains: Report of the Great Plains Committee.* Washington, DC: GPO, 1936.

Hales, Peter Bacon. *Atomic Spaces: Living on the Manhattan Project.* Urbana: Univ. of Illinois Press, 1997.

Hamilton, Shane. "Agribusiness, the Family Farm, and the Politics of Technological Determinism in the Post-World War II United States." *Technology and Culture* 55, no. 3 (July 2014): 560–90.

Hammerman, Donald Robert. "A Historical Analysis of the Socio-Cultural Factors That Influenced the Development of Camping Education." EdD diss., Pennsylvania State Univ., 1961.

Hardin, Charles M. *The Politics of Agriculture: Soil Conservation and the Struggle for Power in Rural America.* Glencoe, IL: Free Press, 1952.

Harmel, R. D., J. V. Bonta, and C. W. Richardson. "The Original USDA-ARS Experimental Watersheds in Texas and Ohio: Contributions from the Past and Visions for the Future." *Transactions of the American Society of Agricultural and Biological Engineers* 50, no. 5 (2007): 1669–75.

Harris, M. Coleman. "A Salute to Jonathan Forman, M.D., F.A.C.A." *Annals of Allergy* 25 (Dec. 1967): 706–10.

Hart, John Fraser. *The Changing Scale of American Agriculture.* Charlottesville: Univ. of Virginia Press, 2003.

Hayden, Cassius C., Albert E. Perkins, C. F. Monroe, William E. Krauss, R. G. Washburn, and C. E. Knoop. *Hay-Crop Silage: A Summary of Ten Years' Work.* Bulletin 656. Wooster: OAES, 1945.

Hayes, William A. *Minimum Tillage Farming.* Brookfield, WI: No-Till Farmer, 1982.

Hays, Samuel P. *Beauty, Health, and Permanence: Environmental Politics in the United States, 1955–1985.* New York: Cambridge Univ. Press, 1987.

———. *Conservation and the Gospel of Efficiency: The Progressive Conservation Movement, 1890–1920.* Cambridge, MA: Harvard Univ. Press, 1959.

Hays, Sandy Miller. "Making the Plant/Soil/Nutrition Connection." *Agricultural Research* 39, no. 5 (1991): 16–21.

Hazlett, Maril Pearce Trigg. "The Story of *Silent Spring* and the Ecological Turn." PhD diss., Univ. of Kansas, 2003.
Headley, Joseph C. "Soil Conservation and Cooperative Extension." In *The History of Soil and Water Conservation,* edited by Douglas Helms and Susan L. Flader, 188–204. Washington, DC: Agricultural History Society, 1985.
Heckman, Joseph R. "A History of Organic Farming: Transitions from Sir Albert Howard's *War in the Soil* to USDA National Organic Program." *Renewable Agriculture and Food Systems* 21, no. 3 (2006): 143–50.
———. "Securing Fresh Food from Fertile Soil, Challenges to the Organic and Raw Milk Movements." *Renewable Agriculture and Food Systems* 34 (2019): 472–85.
———. "Soil Fertility Management a Century Ago in *Farmers of Forty Centuries.*" *Sustainability* 5 (2013): 2795–801.
Helms, Douglas. "Early Leaders of the Soil Survey." In *Profiles in the History of the U.S. Soil Survey,* edited by Douglas Helms, Anne B. W. Effland, and Patricia J. Durana, 20–28. Ames: Iowa State Press, 2002.
———. *Hugh Hammond Bennett and the Creation of the Soil Conservation Service, September 19, 1933—April 27, 1935.* Historical Insights No. 9. Washington, DC: NRCS, 2010.
———. *Hugh Hammond Bennett and the Creation of the Soil Erosion Service.* Historical Insights No. 8. Washington, DC: NRCS, 2008.
———. *Hydrologic and Hydraulic Research in the Soil Conservation Service,* USDA NRCS Historical Insights No. 7. Washington, DC: NRCS, 2007.
———. *Readings in the History of the Soil Conservation Service.* SCS Historical Notes No. 1. Washington, DC: GPO, 1992.
Henderson, Henry L., and David B. Woolner, eds. *FDR and the Environment.* New York: Palgrave Macmillan, 2005.
Hendrickson, Roy F. *Food "Crisis."* Garden City, NY: Doubleday, Doran, 1943.
Herrington, Barbour L. *Milk and Milk Processing.* New York: McGraw-Hill, 1948.
Hersey, Mark D. *My Work Is That of Conservation: An Environmental Biography of George Washington Carver.* Athens: Univ. of Georgia Press, 2011.
Hersey, Mark D., and Jeremy Vetter. "Shared Ground: Between Environmental History and the History of Science." *History of Science* 57, no. 4 (2019): 403–40.
Heyman, Stephen. *The Planter of Modern Life: Louis Bromfield and the Seeds of a Food Revolution.* New York: W. W. Norton, 2020.
Hildebrand, Peter E. "Agronomy's Role in Sustainable Agriculture: Integrated Farming Systems." *Journal of Production Agriculture* 3, no. 3 (1990): 285–88.
Hilts, Philip J. *Protecting America's Health: The FDA, Business, and One Hundred Years of Regulation.* New York: Alfred A. Knopf, 2003.
Hoff, Derek S. *The State and the Stork: The Population Debate and Policy Making in US History.* Chicago: Univ. of Chicago Press, 2012.
Holbrook, Stewart H. *Machines of Plenty: Pioneering in American Agriculture.* New York: Macmillan, 1955.
Hone, Elizabeth B. "An Analysis of Conservation Education in Curriculums for Grades Kindergarten through Twelve." EdD diss., Univ. of Southern California, 1959.
Hopkins, Cyril G. *Soil Fertility and Permanent Agriculture.* Boston: Ginn, 1910.
Howard, Albert. *An Agricultural Testament.* London: Oxford Univ. Press, 1940.
———. *Farming and Gardening for Health or Disease.* London: Faber & Faber, 1945.
———. *The War in the Soil.* Emmaus, PA: Rodale, 1946.
Howard, Albert, and Yeshwant D. Wad. *The Waste Products of Agriculture: Their Utilization as Humus.* 1931, Reprint, Oxford City Press, 2011.

Huber, Peter. *Hard Green: Saving the Environment from the Environmentalists: A Conservative Manifesto.* New York: Basic Books, 1999.

Huffman, C. F., and C. W. Duncan. "The Nutritional Deficiencies in Farm Mammals on Natural Feeds." *Annual Reviews of Biochemistry* 13 (1944): 467–86.

Hurt, R. Douglas. *The Dust Bowl: An Agricultural and Social History.* Chicago: Nelson-Hall, 1981.

———. *Problems of Plenty: The American Farmer in the Twentieth Century.* Chicago: Ivan R. Dee, 2002.

Hutchinson, H. B., and E. H. Richards. "Artificial Farmyard Manure." *Journal of the Ministry of Agriculture* 28, no. 5 (Aug. 1921): 398–411.

Huxley, Julian. *TVA: Adventure in Planning.* Surrey: Architectural Press, 1943.

Igler, David. "On Vital Areas, Categories, and New Opportunities." *Journal of American History* 100, no. 1 (June 2013): 120–23.

Iowa State Univ. Center for Agricultural Adjustment, ed. *Problems and Policies of American Agriculture.* Ames: Iowa State Univ. Press, 1959.

Iversen, Kenneth W. "A Critical Evaluation of Claims Regarding the Relationship between Soil Composition and Human Health." PhD diss., New York Univ., 1956.

Jackson, Carlton. *J. I. Rodale: Apostle of Nonconformity.* New York: Pyramid Books, 1974.

Jacobs, Meg. *Panic at the Pump: The Energy Crisis and the Transformation of American Politics in the 1970s.* New York: Hill & Wang, 2016.

Jacobson, Robert E., and Kent F. Hoddick. *Adjustments by Dairy Marketing Cooperatives in the North Central States.* North Central Regional Research Publication 212. Research Bulletin 1054. Wooster: OARDC, 1972.

Johnson, Carl S. "Conservation Education in Ohio." *Conservation Volunteer* 14, no. 80 (Jan.–Feb. 1951): 39–44.

Johnson, Charles W., and Charles O. Jackson. *City behind a Fence: Oak Ridge, Tennessee, 1942–1946.* Knoxville: Univ. of Tennessee Press, 1981.

Johnston, A. E., and P. R. Poulton. "The Importance of Long-Term Experiments in Agriculture: Their Management to Ensure Continued Crop Production and Soil Fertility; the Rothamsted Experience." *European Journal of Soil Science* 69 (Jan. 2018): 113–25.

Jones, Coleen M., ed. *One Hundred Years of Inquiry and Innovation: An Illustrated History of the American Dairy Science Association.* Long Prairie, MN: Banta, 2006.

Jundt, Thomas. *Greening the Red, White, and Blue: The Bomb, Big Business, and Consumer Resistance in Postwar America.* New York: Oxford Univ. Press, 2014.

Judd, Richard W. *Common Lands, Common People: The Origins of Conservation in Northern New England.* Cambridge, MA: Harvard Univ. Press, 1997.

———. *The Untilled Garden: Natural History and the Spirit of Conservation in America, 1740–1840.* New York: Cambridge Univ. Press, 2009.

Kaikow, Julius. "The Legal and Administration Status of Conservation Education in the United States." PhD diss., Columbia Univ., 1954.

Kallet, Arthur, and F. J. Schlink. *100,000,000 Guinea Pigs: Dangers in Everyday Foods, Drugs, and Cosmetics.* New York: Grosset & Dunlap, 1933.

Keeney, Dennis. "Sustainable Agriculture: Definitions and Concepts." *Journal of Production Agriculture* 3, no. 3 (1990): 281–85.

Kellogg, Charles E. "Food Production Potentialities and Problems." *Journal of Farm Economics* 31, no. 1 (1949): 251–62.

Kelsey, Darwin P. "Outdoor Museums and Historical Agriculture." *Agricultural History* 46, no. 1 (Jan. 1972): 105–28.

King, Franklin H. *Farmers of Forty Centuries: Organic Farming in China, Korea, and Japan.* 1911. Reprint, Mineola, NY: Dover, 2004.

King, Judson. *The Conservation Fight: From Theodore Roosevelt to the Tennessee Valley Authority.* Washington, DC: Public Affairs Press, 1959.

Kline, Benjamin. *First Along the River: A Brief History of the U.S. Environmental Movement,* 3rd ed. Lanham, MA: Rowman & Littlefield, 2007.

Korcak, Ronald F. "Early Roots of the Organic Movement: A Plant Nutrition Perspective." *HortTechnology* 2, no. 2 (Apr.–June 1992): 263–67.

Krech, Shepard, III. *The Ecological Indian: Myth and History.* New York: W. W. Norton, 1999.

Kyle, John H. *The Building of TVA: An Illustrated History.* Baton Rouge: Louisiana State Univ. Press, 1958.

Lacy, Leslie Alexander. *The Soil Soldiers: The Civilian Conservation Corps in the Great Depression.* Radnor, PA: Chilton, 1976.

Lal, Rattan, and J. P. Bruce. "The Potential of World Cropland Soils to Sequester C and Mitigate the Greenhouse Effect." *Environmental Science and Policy* 2 (1999): 177–85.

Lal, Rattan, C. C. Reicosky, and J. D. Hanson. "Evolution of the Plow over 10,000 Years and the Rationale for No-Till Farming." *Soil and Tillage Research* 93, no. 1 (2007): 1–12.

Lamb, Ruth deForest. *American Chamber of Horrors: The Truth about Food and Drugs.* New York: Farrar & Rinehart, 1936.

Lauber, John. "And It Never Needs Painting: The Development of Residential Aluminum Siding." *APT Bulletin: The Journal of Preservation Technology* 31, nos. 2–3 (2000): 19–21.

Lear, Linda. *Rachel Carson: Witness for Nature.* New York: Henry Holt, 1997.

Leopold, Aldo. *A Sand County Almanac: With Essays on Conservation from Round River.* New York: Ballantine, 1970.

Liebig, Justus von. *Chemistry in Its Applications to Agriculture and Physiology, From the Fourth London Edition, Revised and Enlarged.* New York: John Wiley, 1872.

———. *The Natural Laws of Husbandry.* Ed. John Blyth. New York: D. Appleton, 1863.

Lilienthal, David E. *TVA: Democracy on the March.* 20th anniversary ed. New York: Harper & Row, 1953.

Lindberg, John. "Food Supply under a Program of Freedom from Want." *Social Research* 12, no. 2 (May 1945): 181–204.

Linnér, Björn-Ola. *The Return of Malthus: Environmentalism and Post-War Population-Resource Crises.* Isle of Harris, UK: White Horse, 2003.

Little, Charles E. *Green Fields Forever: The Conservation Tillage Revolution in America.* Washington, DC: Island Press, 1987.

Lord, Russell. *Behold Our Land.* Boston: Houghton Mifflin, 1938.

———. *The Care of the Earth: A History of Husbandry.* New York: Mentor Books, 1962.

———. *To Hold This Soil.* USDA Miscellaneous Publication 321. Washington, DC: GPO, 1938.

Lord, Russell, and Kate Lord. *Forever the Land: A Country Chronicle and Anthology.* New York: Harper & Brothers, 1950.

Lowdermilk, Walter C. "Ecological Principles: W. C. Lowdermilk Offers Basic Approach to Sustained Land Use." *Soil Conservation* 3, no. 3 (Sept. 1937): 54, 99.

Lyons, Chuck. "Harnessing Power: Ferguson's 3-Point Hitch Remains an Enduring Milestone in Engineering." *Farm Collector* (Jan. 2018): 30–31.

MacRae, Rod J., Stuart B. Hill, John Henning and Guy R. Mehuys. "Agricultural Science and Sustainable Agriculture: A Review of the Existing Scientific Barriers to Sustainable Food Production and Potential Solutions." *Biological Agriculture and Horticulture* 6, no. 3 (1989): 173–219.

Maher, Neil M. *Nature's New Deal: The Civilian Conservation Corps and the Roots of the American Environmental Movement.* New York: Oxford Univ. Press, 2005.

Malthus, Thomas Robert. *An Essay on the Principle of Population: The 1803 Edition.* Edited by Shannon C. Stimson. New Haven: Yale Univ. Press, 2018.

Manlay, Raphaël J., Christian Feller, and M. J. Swift. "Historical Evolution of Soil Organic Matter Concepts and Their Relationships with the Fertility and Sustainability of Cropping Systems." *Agriculture, Ecosystems and Environment* 119 (2007): 217–33.

Marsh, George P. *The Earth as Modified by Human Action: A New Edition of Man and Nature.* New York: Scribner, Armstrong, 1874.

Maynard, Leonard A. "Effect of Fertilizers on the Nutritional Value of Foods." *Journal of the American Medical Association* 161, no. 15 (Aug. 11, 1956): 1478–83.

McBride, Charles G. *The Ohio Farmer and His Milk Market.* Bulletin 614. Wooster: OAES, 1940.

McCarrison, Robert. *Nutrition and Health: Being the Cantor Lectures Delivered before the Royal Society of Arts, 1936, Together with Two Earlier Essays.* London: Faber & Faber, 1961.

———. *The Work of Sir Robert McCarrison.* Edited by H. M. Sinclair. London: Faber & Faber, 1953.

McCollum, Elmer V. "Relationship between Diet and Dental Caries." *Journal of Dental Research* 11, no. 4 (Aug. 1931): 553–71.

McConnell, Grant. "The Conservation Movement—Past and Present." *Western Political Quarterly* 7, no. 3 (1954): 463–78.

McDonald, Angus. *Early American Soil Conservationists.* USDA Miscellaneous Publication 449. Washington, DC: GPO, 1941.

McIntosh, Robert P. "H. A. Gleason—'Individualistic Ecologist' 1882–1975: His Contributions to Ecological Theory." In *History of American Ecology,* edited by Frank N. Egerton. New York: Arno, 1977.

———. *The Background of Ecology: Concept and Theory.* New York: Cambridge Univ. Press, 1985.

McLean, Eugene O., G. E. Smith, and William A. Albrecht. "Biological Assays of Some Soil Types Under Treatments." *Soil Science Society Proceedings* 8 (1944): 282–86.

McMurry, Sally. "The Impact of Sanitation Reform on the Farm Landscape in U.S. Dairying, 1890–1950." *Buildings & Landscapes: Journal of the Vernacular Architecture Forum* 20, no. 2 (Fall 2013): 22–47.

McNeill, John R. "Observations on the Nature and Culture of Environmental History." *History and Theory* 42 (Dec. 2003): 5–43.

McWilliams, James E. *American Pests: The Losing War on Insects from Colonial Times to DDT.* New York: Columbia Univ. Press, 2008.

Meadows, Donnella H., Dennis L. Meadows, Jørgen Randers, and William W. Behrens III. *The Limits to Growth: A Report for the Club of Rome's Project on the Predicament of Mankind.* New York: Universe Books, 1972.

Meine, Curt. *Aldo Leopold: His Life and Work.* Madison: Univ. of Wisconsin Press, 2010.

Merrill, Karen R., ed. *The Oil Crisis of 1973–1974: A Brief History with Documents.* Boston: Bedford / St. Martin's, 2007.

Merrill, Margaret C. "Eco-Agriculture: A Review of Its History and Philosophy." *Biological Agriculture & Horticulture* 1, no. 3 (1983): 181–210.

Michigan State University. *Nutrition of Plants, Animals, Man.* Michigan State University Centennial Symposium, sponsored by College of Agriculture. East Lansing: Michigan State Univ., 1955.

Miller, Char. *Gifford Pinchot and the Making of Modern Environmentalism.* Washington, DC: Island Press, 2001.

———. *Gifford Pinchot: The Evolution of an American Conservationist.* Milford, PA: Grey Towers Press, 1992.

Miraglia, Laurie L. "Seeds of Knowledge: The Evolution of the Louis Bromfield Sustainable Agriculture Library." MA thesis, Kent State Univ., 1994.

Mitchell, Glen H., and Elmer F. Baumer. *An Interim Report on Milk Marketing Distribution Systems in Ohio.* Research Circular 29. Wooster: OAES, 1956.

Morgan, Robert J. *Governing Soil Conservation: Thirty Years of the New Decentralization.* Baltimore, MD: Johns Hopkins Univ. Press, 1965.

Morison, Francis L. *A Study of the Newer Hay Harvesting Methods on Ohio Farms.* Bulletin 636. Wooster: OAES, 1942.

Mt. Pleasant, Jane. "The Paradox of Plows and Productivity: An Agronomic Comparison of Cereal Grain Production under Iroquois Hoe Culture and European Plow Culture in the Seventeenth and Eighteenth Centuries." *Agricultural History* 85, no. 4 (Fall 2011): 460–92.

Munoz, Samuel E., David J. Mladenoff, Sissel Schroeder, and John W. Williams. "Defining the Spatial Patterns of Historical Land Use Associated with the Indigenous Societies of Eastern North America." *Journal of Biogeography* 41, no. 12 (Dec. 2014): 2195–210.

Nash, Roderick. *Wilderness and the American Mind,* 3rd ed. New Haven: Yale Univ. Press, 1982.

National Wildlife Federation. *Conference on Education in Conservation: Held at the Annual Meeting of the National Wildlife Federation, February 16, 1939, Detroit, Michigan.* Detroit: National Wildlife Federation, 1939.

———. *Education in Conservation: A Report of Progress and Round-Table Discussion, Held at the Fifth North American Wildlife Conference, March 18, 1940, Washington, D. C.* Washington, DC: National Wildlife Federation, 1940.

Nearing, Helen, and Scott Nearing. *Living the Good Life: How to Live Sanely and Simply in a Troubled World.* 1954. Reprint, New York: Schocken, 1970.

Nelson, Charles S. "Proceedings of the Council." *Ohio State Medical Journal* 54, no. 11 (Nov. 1958): 1467–76.

Nelson, Lewis B. *History of the U.S. Fertilizer Industry.* Muscle Shoals, AL: Tennessee Valley Authority, 1990.

Nelson, Robert H. *The New Holy Wars: Economic Religion vs. Environmental Religion in Contemporary America.* University Park: Pennsylvania State Univ. Press, 2010.

Neufeld, Edward P. *A Global Corporation: A History of the International Development of Massey-Ferguson Limited.* Toronto: Univ. of Toronto Press, 1969.

Noble Foundation. *Tracing Our Steps: A History of the Samuel Roberts Noble Foundation.* Ardmore, OK: Noble Foundation, 1995.

Nott, Sherrill B. *Evolution of Dairy Grazing in the 1990s.* Department of Agricultural Economics Staff Paper 2003–07. East Lansing: Michigan State Univ., June 2003.

Odum, Eugene P. *Fundamentals of Ecology.* Philadelphia: W. B. Saunders, 1953.

———. "The New Ecology." *BioScience* 14, no. 7 (July 1964): 14–16.

Olmstead, Alan L., and Paul W. Rhode. *Creating Abundance: Biological Innovation and American Agricultural Development.* New York: Cambridge Univ. Press, 2008.
Osborn, Fairfield. *Our Plundered Planet.* Boston: Little, Brown, 1948.
O'Sullivan, Robin. *American Organic: A Cultural History of Farming, Gardening, Shopping, and Eating.* Lawrence: Univ. Press of Kansas, 2015.
Overholt, James. *These Are Our Voices: The Story of Oak Ridge 1942–1970.* Oak Ridge, TN: Children's Museum of Oak Ridge, 1987.
Palmer, E. Laurence. "Conservation Education." *National Association of Secondary-School Principals: Special Problems in High School Science Education* (Jan. 1953): 191–98.
Perkins, John H. *Geopolitics and the Green Revolution: Wheat, Genes, and the Cold War.* New York: Oxford Univ. Press, 1997.
Petersen, William E., and Albert M. Field. *Dairy Farming.* Chicago: J. B. Lippincott, 1953.
Petulla, Joseph M. *American Environmental History: The Exploitation and Conservation of Natural Resources.* San Francisco: Boyd & Fraser, 1977.
Phillips, Sarah T. *This Land, This Nation: Conservation, Rural America, and the New Deal.* New York: Cambridge Univ. Press, 2007.
Pinchot, Gifford. *Breaking New Ground.* New York: Harcourt, Brace, 1947.
———. *The Fight for Conservation.* New York: Doubleday, Page, 1910.
Pinkett, Harold T. *Gifford Pinchot: Private and Public Forester.* Urbana: Univ. of Illinois Press, 1970.
Pirtle, Thomas R. *History of the Dairy Industry.* Chicago: Mojonnier, 1926.
Pollan, Michael. *The Omnivore's Dilemma: A Natural History of Four Meals.* New York: Penguin, 2006.
Pratt, Avery D., Roy G. Washburn, and Charles F. Rogers. *Comparative Palatabilities of Silages.* Research Bulletin 814. Wooster: OAES, 1958.
Pratt, Avery D., R. R. Davis, H. R. Conrad, and J. H. Vandersall. *Soilage and Silage for Milk Production.* Research Bulletin 871. Wooster: OAES, 1961.
Price, Weston A. *Nutrition and Physical Degeneration: A Comparison of Primitive and Modern Diets and Their Effects,* 4th ed. Redlands, CA: Published by the author, 1945.
Ramser, John H., Frank W. Andrew, and Robert W. Kleis. *Better Hay by Forced-Air Drying.* Univ. of Illinois Extension Service in Agriculture and Home Economics Circular 757. Urbana: Univ. of Illinois, 1956.
Rawlinson, Fred. *Make Mine Milk: The Story of the Milk Bottle and the History of Dairying.* Newport News, VA: Far Publications, 1970.
Rehm, George. *Twelve Cows—And We're in Clover: The Story of a Man Who Bought a Farm.* New York: William Morrow, 1951.
Reilly, Gretchen Ann. "'This Poisoning of Our Drinking Water': The American Fluoridation Controversy in Historical Context, 1950–1990." PhD diss., George Washington Univ., 2001.
Riedman, Sarah R. *Grass: Our Greatest Crop.* New York: Thomas Nelson & Sons, 1952.
Renner, Martin. "Conservative Nutrition: The Industrial Food Supply and Its Critics, 1915–1985." PhD diss., Univ. of California Santa Cruz, 2012.
Robertson, Thomas. *The Malthusian Moment: Global Population Growth and the Birth of American Environmentalism.* New Brunswick, NJ: Rutgers Univ. Press, 2012.
Robinson, George O. *The Oak Ridge Story: The Saga of a People Who Share in History.* Kingsport, TN: Southern Publishers, 1950.
Rodale, J. I. *The Organic Front.* Emmaus, PA: Rodale Press, 1948.
———. *Organic Merry-Go-Round.* Emmaus, PA: Rodale Press, 1954.

———. *Pay Dirt: Farming and Gardening with Composts.* Emmaus, PA: Rodale Press, 1945.
Rothman, Hal K. *The Greening of a Nation? Environmentalism in the United States since 1945.* Fort Worth, TX: Harcourt Brace, 1998.
Russell, Edmund. *War and Nature: Fighting Humans and Insects with Chemicals from World War I to Silent Spring.* New York: Cambridge Univ. Press, 2001.
Russell, E. John. "Rothamsted and Agricultural Science." *Nature* 111, no. 2788 (Apr. 7, 1923): 466–70.
———. "Rothamsted and Its Experiment Station." *Agricultural History* 16, no. 4 (1942): 161–83.
Ruuskanen, Esa, and Kari Väyrynen. "Theory and Prospects of Environmental History." *Rethinking History* 21, no. 4 (2017): 456–73.
Ruxin, Joshua Nalibow. "Hunger, Science, and Politics: FAO, WHO, and Unicef Nutrition Policies, 1945–1978." PhD diss., Univ. College London, 1996.
Rymsza-Pawlowska, M. J. *History Comes Alive: Public History and Popular Culture in the 1970s.* Chapel Hill: Univ. of North Carolina Press, 2017.
Salter, Robert M. "World Soil and Fertilizer Resources in Relation to Food Needs." *Science,* n.s., 105, no. 2734 (May 23, 1947): 533–38.
Sampson, R. Neil. *Farmland or Wasteland: A Time to Choose; Overcoming the Threat to America's Farm and Food Future.* Emmaus, PA: Rodale, 1981.
———. *For Love of the Land: A History of the National Association of Conservation Districts.* League City, TX: National Association of Conservation Districts, 1985.
Schlebecker, John T. *A History of American Dairying.* Chicago: Rand McNally, 1967.
———. *Whereby We Thrive: A History of American Farming, 1607–1972.* Ames: Iowa State Univ. Press, 1975.
Schultz, Stanley R., and Elmer F. Baumer. *Shifts in the Ohio Dairy Industry.* Research Circular 80. Wooster: OAES, 1959.
Schutz, Willard M. "First Place Winner: We Must . . . Put Something Back." *Plant Food Journal* 3, no. 4 (Oct.–Dec. 1949): 4.
Schweitzer, Albert. *Out of My Life and Thought: An Autobiography.* Baltimore: Johns Hopkins Univ. Press, 1990.
Scott, Ivan. *Louis Bromfield, Novelist and Agrarian Reformer: The Forgotten Author.* Lewiston, NY: Edwin Mellen, 1998.
Sears, Paul B. *Deserts on the March.* Norman: Univ. of Oklahoma Press, 1935.
———. *The Ecology of Man.* Eugene: Oregon State System of Higher Education, 1957.
———. "Human Ecology: A Problem in Synthesis." *Science,* n.s. 120, no. 3128 (Dec. 10, 1954): 959–63.
———. "Soil and Health." *Ohio State Medical Journal* 44, no. 3 (Mar. 1948): 269–71.
———. *This Is Our World.* Norman: Univ. of Oklahoma Press, 1937.
Sears, Paul B., and George F. Carter. "Ecology and the Social Sciences: A Reply." *Ecology* 33, no. 2 (Apr. 1952): 299–300.
Serviss, George H., and Gilbert H. Ahlgren. *Grassland Farming.* New York: John Wiley & Sons, 1955.
Shabecoff, Philip. *A Fierce Green Fire: The American Environmental Movement.* Washington, DC: Island Press, 2003.
Shaudys, Edgar T. "A Critical Analysis of Ohio Dairy Production Cost Studies, with Special Emphasis on the Farm Approach." PhD diss., Ohio State Univ., 1954.
Shaudys, Edgar T., John H. Sitterley, and J. A. Studebaker. *Costs of Storing Grass-Legume Silage.* Research Bulletin 853. Wooster: OAES, June 1960.

Shaw, D. John. *World Food Security: A History since 1945.* New York: Palgrave Macmillan, 2007.
Shepard, Paul, and Daniel McKinley, eds. *The Subversive Science: Essays toward an Ecology of Man.* Boston: Houghton Mifflin, 1969.
Shulman, Stuart W. "The Business of Soil Fertility: A Convergence of Urban-Agrarian Concern in the Early 20th Century." *Organization & Environment* 12, no. 4 (1999): 401–24.
Sinnott, Edmund W. "Paul B. Sears." *Science,* n.s., 121, no. 3138 (Feb. 18, 1955): 227.
Smil, Vaclav. *Enriching the Earth: Fritz Haber, Carl Bosch, and the Transformation of World Food Production.* Cambridge: MIT Press, 2001.
Smith, Frederick E. "Ecology and the Social Sciences." *Ecology* 32, no. 4 (Oct. 1951): 763–64.
Smith, G. E., and William A. Albrecht. "Feed Efficiency in Terms of Biological Assays of Soil Treatments." *Soil Science Society Proceedings* 7 (1943): 322–30.
Smith-Howard, Kendra. *Pure and Modern Milk: An Environmental History since 1900.* New York: Oxford Univ. Press, 2014.
Smith, Robert Leo, ed. *The Ecology of Man: An Ecosystem Approach.* New York: Harper & Row, 1972.
Snow, Whitney A. "A Great Dream for This Valley: Louis Bromfield and Wichita Falls Malabar Farm, 1949–1954." *Southwestern Historical Quarterly* 119, no. 4 (Apr. 2016): 378–405.
Souder, William. *On a Farther Shore: The Life and Legacy of Rachel Carson.* New York: Crown, 2012.
Staples, Amy L. S. *The Birth of Development: How the World Bank, Food and Agriculture Organization, and World Health Organization Changed the World, 1945–1965.* Kent, OH: Kent State Univ. Press, 2006.
Staten, Hi W. *Grasses and Grassland Farming.* New York: Devon-Adair, 1952.
Steinberg, Ted. "Down to Earth: Nature, Agency, and Power in History." *American Historical Review* (June 2002): 798–820.
Stoll, Steven. *Larding the Lean Earth: Soil and Society in Nineteenth-Century America.* New York: Hill & Wang, 2002.
Stradling, David, ed. *Conservation in the Progressive Era: Classic Texts.* Seattle: Univ. of Washington Press, 2004.
———, ed. *The Environmental Moment: 1968–1972.* Seattle: Univ. of Washington Press, 2012.
Stuckey, Ronald L. "Paul Bigelow Sears (1891–1990): Eminent Scholar, Ecologist and Conservationist." *Ohio Journal of Science* 109, no. 4–5 (2010): 140–44.
Sutter, Paul S. *Let Us Now Praise Famous Gullies: Providence Canyon and the Soils of the South.* Athens: Univ. of Georgia Press, 2015.
———. "The World with Us: The State of American Environmental History." *Journal of American History* 100, no. 3 (June 2013): 94–119.
Swanson, Drew A. *A Golden Weed: Tobacco and Environment in the Piedmont South.* New Haven: Yale Univ. Press, 2014.
Tanner, C. B., and R. W. Simonson. "Franklin Hiram King—Pioneer Scientist." *Soil Science Society of America Journal* 57 (Jan.–Feb. 1993): 286–92.
Tansley, Arthur G. "The Use and Abuse of Vegetational Concepts and Terms." *Ecology* 16, no. 3 (July 1935): 284–307.
Terry, Cyrl Waldie. "The Thermodynamics of Hay Driers." PhD diss., Cornell Univ., 1948.

Thornthwaite, Charles W. "The Research Program of the Section of Climatic and Physiographic Research." *Soil Conservation* 2, no. 10 (Apr. 1937): 218–20, 236.

Tobey, Ronald C. *Saving the Prairies: The Life Cycle of the Founding School of American Plant Ecology, 1895–1955*. Berkeley: Univ. of California Press, 1981.

Tolley, Howard R. "Farmers in a Hungry World." *Proceedings of the American Philosophical Society* 95, no. 1 (Feb. 13, 1951): 54–61.

Tolley, Howard R., F. F. Elliott, J. F. Thackrey, F. V. Waugh, and Rainer Schickele. "Agriculture in the Transition from War to Peace." *American Economic Review* 32, no. 2 (May 1945): 390–404.

Treadwell, Danielle D., D. E. McKinney, and Nancy G. Creamer. "From Philosophy to Science: A Brief History of Organic Horticulture in the United States." *HortScience* 38, no. 5 (Aug. 2003): 1009–14.

Tyrell, Ian. *Crisis of the Wasteful Nation: Empire and Conservation in Theodore Roosevelt's America.* Chicago: Univ. of Chicago Press, 2015.

Udall, Stewart L. *The Quiet Crisis and the Next Generation.* Layton, UT: Gibbs Smith, 1988.

USDA. *Agricultural Land Requirements and Resources: Part III of the Report on Land Planning.* Washington, DC: GPO, 1935.

———. *Agricultural Statistics 1952*. Washington, DC: GPO, 1952.

———. *The Effect of Soils and Fertilizers on the Nutritional Quality of Plants.* Agriculture Information Bulletin No. 299. Washington, DC: GPO, 1965.

———. *Factors Affecting the Nutritive Value of Foods: Studies at the U.S. Plant, Soil, and Nutrition Laboratory.* USDA miscellaneous publication 664. Washington, DC: GPO, 1948.

———. *Grass: The Yearbook of Agriculture, 1948*. Washington, DC: GPO, 1948.

———. *Report and Recommendations on Organic Farming: Prepared by USDA Study Team on Organic Farming.* Washington, DC: USDA, July 1980.

———. *Save Fuel . . . Use Conservation Tillage.* SCS Program Aid 1263. USDA: May 1980.

———. *Soils and Men: Yearbook of Agriculture 1938*. Washington, DC: GPO, 1938.

———. *Summary Report: 2012 National Resources Inventory.* Washington, DC: NRCS, 2015.

———. *Summary Report: 2015 National Resources Inventory.* Washington, DC: NRCS, 2018.

———. *Tillage Options for Conservation Farmers.* Program Aid 1416. Washington, DC: USDA, 1989.

Valenze, Deborah. *Milk: A Local and Global History.* New Haven: Yale Univ. Press, 2011.

Van der Ploeg, R. R., W. Böhm, and M. B. Kirkham. "History of Soil Science: On the Origin of the Theory of Mineral Nutrition of Plants and the Law of the Minimum." *Soil Science Society of America Journal* 63 (1999): 1055– 62.

Vogt, William. *Road to Survival.* New York: William Sloane, 1948.

Waksman, Selman A. *Humus: Origin, Chemical Composition, and Importance in Nature*, 2nd ed. Baltimore: Williams & Wilkins, 1938.

Waksman, Selman A., Florence G. Tenney, and Robert A. Diehm. "Chemical and Microbiological Properties Underlying the Transformation of Organic Matter in the Preparation of Artificial Manures." *Journal of the American Society of Agronomy* 21, no. 5 (1929): 533–46.

Ward, Henry B., Paul B. Sears, and Cyril J. Ballam, eds. *The Foundations of Conservation Education.* National Wildlife Federation, 1941.

Weber, Margaret B. "Manufacturing the American Way of Farming: Agriculture, Agribusiness, and Marketing in the Postwar Period." PhD diss., Iowa State Univ., 2018.
Wedin, Walter F., and Steven L. Fales, eds. *Grassland: Quietness and Strength for a New American Agriculture.* Madison, WI: American Society of Agronomy, Crop Science Society of America, and Soil Science Society of America, 2009.
Weimar, Mark R., and Don P. Blayney. *Landmarks in the U.S. Dairy Industry.* USDA ERS Agriculture Information Bulletin 694. Washington, DC: USDA, 1994.
Wellum, Caleb. "Energizing the Right: Economy, Ecology, and Culture in the 1970s US Energy Crisis." PhD diss., Univ. of Toronto, 2017.
Wengert, Norman I. *Valley of Tomorrow: The TVA and Agriculture.* Knoxville: Univ. of Tennessee, July 1952.
White, Suzanne Rebecca. "Chemistry and Controversy: Regulating the Use of Chemicals in Foods, 1883–1959." PhD diss., Emory Univ., 1994.
Whitfield, Charles J. "Ecological Relations of Vegetation and Erosion." *Soil Conservation* 3, no. 11 (May 1938): 262–64, 270.
Whitney, Milton. *Soils of the United States, Based upon the Work of the Bureau of Soils to January 1, 1908.* USDA Bureau of Soils Bulletin 55. Washington, DC: GPO, 1909.
Whorton, James. *Before Silent Spring: Pesticides and Public Health in Pre-DDT America.* Princeton, NJ: Princeton Univ. Press, 1974.
Wickenden, Leonard. *Our Daily Poison: The Effects of DDT, Fluorine, Hormones and Other Chemicals on Modern Man.* New York: Devin-Adair, 1955.
Wiley, Harvey W. *Harvey W. Wiley: An Autobiography.* Indianapolis, IN: Bobbs-Merrill, 1930.
Williams, Michael. *Ford and Fordson Tractors.* Poole, UK: Blandford, 1985.
———. *Massey-Ferguson Tractors.* Ipswich, UK: Farming Press, 1987.
Wilson, Louis H. "Conservation of our Soil Resources: Announcing a Nation-Wide Essay Contest Sponsored by the National Grange and American Plant Food Council." *Plant Food Journal* 3, no. 1 (Jan.–Mar. 1949): 2–3.
Witt, Jon. *SOC: Rural Sociology.* Boston: McGraw Hill, 2011.
Woodhouse, Keith Makoto. *The Ecocentrists: A History of Radical Environmentalism.* New York: Columbia Univ. Press, 2018.
Woods, Thomas A. "Living Historical Farming: A Critical Method for Historical Research and Teaching about Rural Life." *Journal of American Culture* 12, no. 2 (Summer 1989): 43–47.
Worster, Donald. *Dust Bowl: The Southern Plains in the 1930s.* New York: Oxford Univ. Press, 1979.
———. *Nature's Economy: A History of Ecological Ideas,* 2nd ed. New York: Cambridge Univ. Press, 1994.
———. "Transformations of the Earth: Toward an Agroecological Perspective in History." *Journal of American History* 76, no. 4 (Mar. 1990): 1087–106.
Wulf, Andrea. *Founding Gardeners: The Revolutionary Generation, Nature, and the Shaping of the American Nation.* New York: Alfred A. Knopf, 2011.
Yoder, Robert E. "Factors Influencing Fertilizer Efficiency." *Plant Food Journal* 2, no. 4 (Apr.–June 1948): 11–13, 31–32.
Young, Gerald L., ed. *Origins of Human Ecology.* Stroudsburg, PA: Hutchinson Ross, 1983.
Young, Harry M., Jr. *No-Tillage Farming.* Brookfield, WI: No-Till Farmer, 1982.
Youngberg, Garth. "The Alternative Agricultural Movement." *Policy Studies Journal* 6, no. 4 (Summer 1978): 524–30.

Youngberg, Garth, and Suzanne P. DeMuth. "Organic Agriculture in the United States: A 30-Year Retrospective." *Renewable Agriculture and Food Systems* 28, no. 4 (2013): 294–328.

Zartman, David L., ed. *Intensive Grazing, Seasonal Dairying: The Mahoning County Dairy Program, 1987–1991*. OARDC Research Bulletin 1190. Wooster, OH: OARDC, 1994.

Zimmerman, Oswald T., and Irvin Lavine. *DDT: Killer of Killers*. Dover, NH: Industrial Research Service, 1946.

# Index

Page numbers in italics refer to illustrations.